Probability
Without Tears

Derek Rowntree

Probability Without Tears

Derek Rowntree

Charles Scribner's Sons · New York

Library of Congress Cataloging in Publication Data

Rowntree, Derek.
　Probability without tears.

　1. Probabilities.　I. Title.
QA273.R8665　　1984　　　519.2　　　84-5510
ISBN 0-684-17994-6
ISBN 0-684-18252-1 (pbk.)

1 3 5 7 9 11 13 15 17 19　FC　20 18 16 14 12 10 8 6 4 2
1 3 5 7 9 11 13 15 17 19　FP　20 18 16 14 12 10 8 6 4 2

PRINTED IN THE UNITED STATES OF AMERICA

Contents

1
Simple Probabilities

Choice and Chance

Because it plays such a big part in our lives, most of us have some idea of what probability is about. We know it involves weighing up the chances or likelihood of something or other taking place. Let's plunge straight into just such a probability situation.

The table shows the wages earned by all the men in a particular factory. ($245 < $250 means 'at least $245 but *less than* $250 per week'; or $245.00 to $249.99.) Thus, we see that 2 men earn at least $245 but less than $250, 7 men earn at least $250 but less than $255, and so on.

Weekly wage $	No. of men
245 < 250	2
250 < 255	7
255 < 260	18
260 < 265	40
265 < 270	21
270 < 275	11
275 < 280	1

Every Christmas the factory holds a raffle in which each man is to have an equal chance of winning. Each man's name is written on a separate plastic disc. The discs are then shaken together in a large box.

How many discs will there be in the box?

There will be 100 discs in the box. (One for each man.)

Before making the draw, we shake the box until the 100 names are well and truly mixed. Then the Plant Manager is blindfolded. He puts his hand into the box and brings out a disc—bearing, of course, the name of the winner.

Ted Brown has been in the raffle for the last seven years but he has never won it. What are his chances this year? Well, there are 100 discs in the box, and only one of them has his name on it. So there is just one chance in 100 that his name will be selected.

What chance does each of the other men have of winning?

Each man has just *one* chance in 100. (Each is equally likely to win.)

Any named individual has just one chance in 100 of winning. But what happens when we consider *groups* of men? For instance, what are the chances that the winner of the raffle will be someone earning between $260 and $265 per week? (Look back at the table.)

Each of the 100 names in the box has an equal chance of being selected. 40 of them belong to men in the $245 < $250 group. So the chances of such a name being chosen are *40 in 100.*

Is the Plant Manager more likely to draw the name of someone in the lowest-paid group ($245 < $250) or someone in the highest-paid group ($275 < $280)? (Look back at the table.)

He is more likely to pick the name of someone in the *lowest*-paid group. (There are two men in the lowest-paid group but only one in the highest-paid group. So there is only one chance in 100 that the chosen name will come from the highest-paid group. But there are two chances in 100 that the chosen name will be from the lowest-paid group.)

Actually, the Plant Manager rather hopes that the prize will go to someone earning less than $255 a week. What are the chances that this will NOT be so? (Look back at the table.)

---

91 chances in 100 (There are 2 + 7 = 9 men earning less than $255 a week, leaving 100 − 9 = 91 men who are earning $255 or more.)

So we have already worked out a number of chances (or probabilities) for the Christmas raffle, e.g.:

Possible result (winner)	Probability
Any particular man	1 in 100
Someone earning $260 < $265	40 in 100
Someone in lowest-paid group	2 in 100
Someone earning not less than $255	91 in 100

Such probabilities or chances are usually written as fractions, e.g.

$$\frac{1}{100} \quad \frac{40}{100} \quad \frac{2}{100} \quad \frac{91}{100} \quad \text{(or 0.01, 0.4, 0.02, 0.91)}$$

Written as a fraction, what are the chances of selecting a man whose pay is at least $270 but less than $275 per week?

The probability is $\frac{11}{100}$ (or 0.11)—11 chances in 100.

Suppose we consider two groups together: What is the probability that the man selected will be earning at least $265 but less than $275?

The probability is $\frac{32}{100}$ (or 0.32). (There are $21 + 11 = 32$ men earning $265 < $275, so they will have 32 chances in the 100.)

For convenience we often 'simplify' fractions like $\frac{32}{100}$.

Thus, $\frac{32}{100}$ would simplify to $\frac{8}{25}$.

So, suppose we now say that the chances of people earning $265 < $275 are $\frac{8}{25}$. Does this mean that their probability of winning is bigger, or smaller, or the same as before?

... the *same* as before. ($\frac{8}{25}$ is the same probability as $\frac{32}{100}$. It's just as good to have 8 chances in 25 as to have 32 in 100, or 16 chances in 50, or 64 in 200, and so on. The *ratio* is what counts.)

Written as a fraction in its simplest form, what is the probability that the name drawn from the box will belong to someone in the $260 < $265 wage-group?

There are 40 in the $260 < $265 group, and 40 chances out of 100 is $\frac{40}{100}$ or $\frac{2}{5}$. Suppose those 40 men in the $260 < $265 wage-group were to *withdraw* their names from the raffle. (Perhaps because they're tired of waiting for us to make the draw!) Suppose their discs were removed from the box and the remainder well-shaken. What is the probability that the Plant Manager would then select the following:
   (i)   Someone in the lowest-paid group?
   (ii) Someone in the highest-paid group?
   (iii) Someone earning $260 < $265?
   (iv) Someone NOT in the $260 < $265 wage-group?

With the 40 members of the $260 < $265 wage-group withdrawn, there are now only 60 names that can be selected. So the probabilities are:

   (i)   2 chances in 60 $= \frac{2}{60} = \frac{1}{30}$

   (ii) 1 chance in 60 $= \frac{1}{60}$

   (iii) 0 chances in 60 $= \frac{0}{60} = 0$ (*Impossible* for member of $260 < $265 group to win)

   (iv) 60 chances in 60 $= \frac{60}{60} = 1$ (*Certain* that someone NOT in $260 < $265 group will win).

And if you want to know who finally won that raffle . . . it was Victor Clasp, the best-paid worker in the factory. Which just goes to show that probability does not favour the poor, not even at Christmas!

Anyway, that simple probability situation has enabled us to bring out a number of points we'll be developing later on. Before we do, I'll give you the opportunity to apply what you've learned to a quite different situation.

It was noticed that some trees in a forest were showing signs of disease. A random* sample of 200 trees of various sizes was examined and the results were as follows:

Type	Disease free	Doubtful	Diseased	Totals
Large	35	18	15	68
Medium	46	32	14	92
Small	24	8	8	40
Totals	105	58	37	200

Calculate the probabilities that:
  (i) one tree selected at random from this sample is both small and diseased;
  (ii) one tree selected at random from this sample is small and either 'doubtful' or disease free;
  (iii) one tree selected at random from the group of medium trees is 'doubtful';
  (iv) one tree selected at random from the complete sample is large.

(i) Of the 200 trees in the sample, only 8 are both small and diseased. So the probability of choosing such a tree is $\frac{8}{200} = \frac{1}{25}$ (or 0.04).

(ii) Of the 200 trees, 8 are small and 'doubtful' while another 24 are small and disease free. So, $8 + 24 = 32$ are small and either 'doubtful' or disease free. Thus, the probability of selecting such a tree from the sample is $\frac{32}{200} = \frac{4}{25}$ (or 0.16).

(iii) Notice that this tree has to be selected from the 'medium' trees—not from among the 200 trees as a whole. Thus we have only 92 chances altogether. Of these 92 trees, 32 are 'doubtful'; so the probability of choosing such a tree is $\frac{32}{92} = \frac{8}{23}$ (or 0.35).

(iv) Here we are back to choosing from the whole sample of 200 trees. 68 of

---

* 'Random' means that each of the trees from among which a choice is made has an *equal* chance of being chosen.

them are large, so the probability of choosing a large one is $\dfrac{68}{200} = \dfrac{17}{50}$ (or 0.34).

So we have seen that it is possible to attach a *number* to the chances of something happening. We can calculate the probability. This we write as a fraction. How do we calculate the probability fraction? Well, the approach we've used so far can be described by using the idea of 'equally likely outcomes'.

Equally Likely Outcomes

To calculate the probability of some result in which we are interested, we can ask.
(a) in how many equally possible ways can the situation turn out?
(b) how many of these equally likely outcomes will give us the result we are looking for?
And we form the probability fraction like this:

$$\text{\textit{Probability of result we are looking for}} = \frac{\text{(b) Number of outcomes giving the 'looked-for' result}}{\text{(a) Total number of equally likely outcomes}}$$

Make a *note* of this formula.

Let's see how this formula applies to our factory raffle. There we had 100 names in the box—100 equally likely outcomes. Suppose the result whose probability we are looking for is 'Drawing the name of someone earning less than $255 per week'. There are 9 such people whose names are in the box. So what is the probability fraction?

$$\frac{\text{No. of outcomes giving 'looked-for' result}}{\text{Total no. of equally likely outcomes}} = \frac{9}{100}$$

Out of 100 equally likely outcomes, 9 will give the result we are looking for. So:

Let's see how the formula applies to other results we might be interested in. Suppose we wanted to know the probability of drawing the name of someone earning less than $240 per week. Here, *none* of the 100 equally likely outcomes gives the looked-for result. (All the men earn at least $245.) So:

$$\frac{\text{No. of outcomes giving looked-for result}}{\text{Total no. of equally likely outcomes}} = \frac{0}{100} = 0$$

On the other hand, what if each and every one of the possible outcomes would give the result we are looking for? Suppose, for instance, that we wanted to know the probability of the factory raffle being won by someone earning *more than* $240 per week.

If *all* the equally likely outcomes would give us the looked-for result, then the probability of that result happening must be __?__.

If *all* the equally likely outcomes would give the looked-for result, then the probability of that result happening must be *1*.

For instance, the probability that the raffle would be won by *someone* working at the factory is:

$$\frac{\text{No. of outcomes giving looked-for result}}{\text{Total no. of equally likely outcomes}} = \frac{100}{100} = 1$$

(That is, all of the 100 factory workers' names are in the raffle, and any one of the 100 names would count as the 'looked-for' result.)

You'll notice that I'm using 'result' as a kind of technical term. It refers to a particular set of *one or more* equally likely outcomes from among all the equally likely outcomes that are possible in a given situation. The 'result', in this sense, is the outcome (or outcomes) we are specially interested in, those we are 'looking for'. (Some textbooks use 'event' instead of 'result' but with the same technical meaning.)

Let's see how this works out in a new situation that has four equally likely outcomes. We'll apply our probability formula to the Porter family (Mom, Dad, Bert and Liz) who are playing a new board-game in which each player has an equal chance of winning.

There are four equally likely outcomes. Let's consider the probability of various possible results. What is the probability that:

(i)   the game will be won? (There's no chance of a draw.)
(ii) the winner will NOT be Bert?
(iii) the winner will be one or other of the two children?
(iv) the winner will be Mom?
(v)   the winner will be the dog, Rover (who is looking on with much interest)?

Using the probability formula:

(i) $\frac{4}{4} = 1$ (ii) $\frac{3}{4}$ (iii) $\frac{2}{4} = \frac{1}{2}$ (iv) $\frac{1}{4}$ (v) $\frac{0}{4} = 0$

A 'Scale' of Probability

In fact, a probability fraction can never be less than 0 (absolute impossibility). Nor can it be greater than 1 (absolute certainty). This is because the number of outcomes giving the result we are looking for can never exceed the total number of possible outcomes. Thus, when we measure probability we are using a 'scale' that runs from ZERO to UNITY:

0 ◄─────────────────── $\frac{1}{2}$ ───────────────────► 1

At which end of the probability scale would you put the probability that
   (i)   one day you will die?
   (ii) I can swim round the world in 30 minutes?

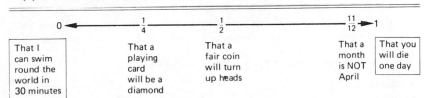

It's certain you will die one day and impossible for me ever to swim round the world. But few of life's possibilities are as clear-cut: rarely does probability equal 0 or 1. Mostly we are dealing with the fractions between 0 and 1—like the probability of $\frac{1}{2}$ that a tossed coin will land 'heads', or of $\frac{1}{4}$ that a randomly-chosen playing card will be a 'diamond'. (See the probability scale.)

In which *half* of the scale would you put the probability that:
   (i)   your entry in the Reader's Digest Sweepstakes will arrive safely by mail?
   (ii) you will win a huge prize?

(i)   The probability of safe arrival must be close to 1.
(ii) The probability of your winning a huge prize must be very close to 0.

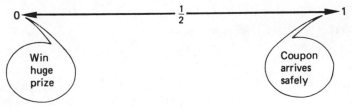

Probability *P*'s and *q*'s

To save space, we often use the letter '*p*' to stand for a given probability. Thus, in talking about my chances of swimming round the world, $p = 0$. But $p = 1$ that you will die some day.

   Suppose I ask you to meet me at a New York area airport, but I forget to tell you which one. You might work out that I could equally well arrive at Kennedy, LaGuardia, or Newark. So what would be your chance of meeting me (*p*) if you went to wait at one of these three airports? In this case, $p = \underline{\quad?\quad}$.

$p = \frac{1}{3}$ (In this case, out of three equally likely outcomes—your waiting at any one of the three airports—only one will give you the looked-for result.)

If '*p*' is the probability of a result happening, the probability of the alternative—failure to happen—is often written as '*q*',

$$p = \text{probability of 'success'}$$
$$q = \text{probability of 'failure'·}$$

In the case of your problem with the three airports, $q = $ ___?___ .

$q = \dfrac{2}{3}$ (There are two outcomes leading to failure and only one that will give you 'success'.)

All right then:

$$p = \text{probability the looked-for result } does\ happen$$
$$q = \text{probability the looked-for result } fails\ to\ happen$$

Since the result we are looking for in any probability 'situation' must *either* happen *or* else fail to happen, the two fractions must add up to 1. That is:

$$p + q = 1·$$

In our last example, $p = \dfrac{1}{3}$, $q = \dfrac{2}{3}$, and $\dfrac{1}{3} + \dfrac{2}{3} = 1·$

Suppose the chances of an astronaut returning safely from a mission to the moon are $p = \dfrac{9}{11}$. What would q mean? And what would be its value?

---

q would be the probability of the astronaut failing to return. Since, in this case, $p = \dfrac{9}{11}$, $q = \dfrac{2}{11}$. (That is, $1 - \dfrac{9}{11}$.)

Note that, if: $p = $ probability of 'success'
and $q = $ probability of 'failure'

$p + q = 1$, therefore $q = 1 - p$ and $p = 1 - q·$

A tourist is walking along a country lane looking for the local church. He comes to a cross-roads and there are five roads to choose from. He has no way of knowing which road will pass the church, all look equally likely.

What is p (probability of passing church) and what is q ? (Check with the diagram.)

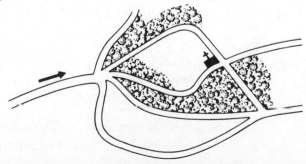

p (probability of success) $= \dfrac{2}{5}$.

q (failure) $= \dfrac{3}{5}$.

(Two roads lead to the church and three do not. So, out of the five possible outcomes, two would give the successful result, three a failure. Notice that, of course, $p + q = 1$.)

If the probability of a particular result happening is:

(i) $\dfrac{3}{4}$ (ii) $\dfrac{5}{8}$ (iii) $\dfrac{1}{4}$ (iv) $\dfrac{1}{2}$ (v) 0.3 (vi) 0.95

what is the probability of its *failing* to happen?

Using your knowledge that $p + q = 1$:

(i) $\dfrac{1}{4}$ (ii) $\dfrac{3}{8}$ (iii) $\dfrac{3}{4}$ (iv) $\dfrac{1}{2}$ (v) 0.7 (vi) 0.05

So we have got used to the idea that probability can be measured on a scale, and we have worked out the probability that a few imaginary results will happen or fail to happen. But how do we actually decide probabilities in real life? There are two main ways—theoretical and practical. We'll look at each in turn.

Theoretical Probability

Sometimes we feel we can state the probability of a result simply by examining the circumstances—'weighing up the possibilities'. For instance, consider the two captains tossing a coin to decide which of their football teams shall kick off. This method is generally thought to be a fair way to begin a game. But what do we mean by 'fair'?

Which of the following is the theory on which we base this idea of fairness?

    (a) Heads on one toss will be followed by tails on the next, and so on?

or (b) The coin is as likely to come down heads as tails?

or (c) In a large number of tosses, the number of heads and tails would be exactly the same?

(b) The coin is as likely to come down heads as tails. The possibility of the coin remaining up in the air or landing on its edge are so remote that only two outcomes remain: heads or tails. Since, *in theory*, there is no reason why a coin should land on one side rather than the other, we say they are both *equally likely* (and the two teams have an equal chance of winning the toss). However, this doesn't mean we should expect *exactly* half and half in a large number of tosses; nor has the coin a memory to enable it to give heads this time if it gave tails last, and vice versa.

It is the idea of equally likely outcomes that allows us to come up with a *theoretical* probability for the coin landing, say, heads. Since this is one of *two* equally likely outcomes, $p(\text{head}) = \frac{1}{2}$.

The probability of the other outcome, $q(\text{tail}) = \underline{\quad?\quad}$.

$q(\text{tail}) = \frac{1}{2}$ (Out of two equally likely outcomes, one gives a head and the other gives a tail.)

If six players had to decide who should start a game, they might do it by rolling a die. (Dice, as you know, have six sides with dots from one to six.) Each player would take a number and the one who rolled that number would start.

What would be the theoretical probability of, say, the player who chose '3 dots' winning the throw?

$p(\text{rolling a 3}) = \frac{1}{6}$ (There are six equally likely outcomes, one of which is the die showing a '3'.)

Now, if $p(\text{rolling a 3}) = \frac{1}{6}$ what is q in this case?

$q(\text{not rolling a 3}) = \frac{5}{6}$ (As always, $p + q = 1$, and since $p = \frac{1}{6}$, $q = \frac{5}{6}$. In other words, out of six equally likely outcomes, only one allows the player choosing '3' to win, while the other five prevent him.)

Suppose our two football captains lost their coin in the mud, and had to use a die instead. How could they use it so as to get *two* equally likely results, and therefore a theoretical probability of $\frac{1}{2}$ that each would win?

Did you see a way? Whether you did or not, look at the list below. Can you see there a pair of results that would give each captain a theoretical probability of $\frac{1}{2}$ of winning?

    (a) '6' or 'any other number'?
  or (b) 'odd number' or 'even number'?
  or (c) 'any number less than 4' or 'any number greater than 4'?

(b) 'odd number' or 'even number' (There are six numbers from 1 to 6 and each is equally likely to be rolled. Three of these numbers are odd, three are even. So odds and evens are equally likely results. The captains could therefore call 'odds' or 'evens' instead of 'heads' or 'tails'.

Of course, any other way of splitting the six possible results into two equally likely results would also have given each captain a theoretical probability of $\frac{1}{2}$, e.g. '1, 2, 3' or '4, 5, 6'. Notice that (a) '6' or 'any other number' would *not* give equally likely results—'any other number' (that is, 1, 2, 3, 4, or 5) is five times as likely as '6'. Similarly (c) 'any number less than 4' or 'any number greater than 4' would not give equally likely results for there are three numbers less than 4 and only two numbers greater.

It is always easy to calculate theoretical probabilities for problems involving coins and dice, for we can easily see what outcomes are equally likely. So it is with playing cards.

From a normal* pack of 52 playing cards, what is the probability of drawing (at random):

(i) a red card?
(ii) a heart?
(iii) a face card?
(iv) a king?
(v) a red king?
(vi) the king of diamonds?

(i) $p = \frac{1}{2}$ (26 of the 52 cards are red)

(ii) $p = \frac{1}{4}$ (13 of the 52 cards are hearts)

(iii) $p = \frac{3}{13}$ (12 of the 52 cards are face cards)

(iv) $p = \frac{1}{13}$ (4 of the 52 cards are kings)

(v) $p = \frac{1}{26}$ (2 of the 52 cards are red kings)

(vi) $p = \frac{1}{52}$ (only 1 of the 52 cards is the king of diamonds)

Now suppose you have *two* chances to draw from the pack of playing cards. What is your theoretical probability of drawing an *ace* (four of them in the pack):

(i) on the first attempt?
(ii) on the second attempt if the first card (which you did NOT put back in the pack)

---

* In case you are not sure . . . a normal pack of 52 cards has four 'suits'—hearts, diamonds, clubs, and spades. Hearts and diamonds are red, while clubs and spades are black. In each suit there are 13 cards: one ace, nine cards numbered 2 to 10, and three 'face' cards (the Jack, Queen, and King). So any card you pick out will be the '7 of hearts' or the 'Queen of clubs', and so on.

(a) was an ace?
(b) was NOT an ace?
(Think carefully over this one.)

(i) $p = \dfrac{1}{13}$ (Since there are 4 aces among 52 cards and $\dfrac{4}{52} = \dfrac{1}{13}$.)

(ii) (a) If the first card chosen *was* an ace, there will be 3 aces only left in the pack. And since one of the cards is now missing from the pack, it holds only 51 cards. So, $p = \dfrac{3}{51}$.

(b) If the first card chosen was NOT an ace, then 4 of the remaining 51 cards will be aces: $p = \dfrac{4}{51}$.

Now try these:
Using the idea of equally likely outcomes, what is the probability of the following results happening:
1 drawing a red ball from a bag containing 4 black balls and 1 red ball?
2 drawing a red ball from a bag containing 8 black balls and 2 red balls?
3 selecting a boy from a class of 20 boys and 16 girls if the pupil is chosen 'at random'?
4 choosing at random a vowel from the word 'probable'?
5 winning a raffle if I have bought 5 of the 200 tickets sold?
6 picking at random an odd number from the numbers 1 to 9?
7 If a row of cinema seats can hold 7 boys and Pete Smith is sitting there with 6 friends, what is the probability that Pete is:
(a) in the middle?
(b) at either end of the row?

---

1 $p = \dfrac{1}{5}$ 2 $p = \dfrac{1}{5}$ 3 $p = \dfrac{20}{36} = \dfrac{5}{9}$ 4 $p = \dfrac{3}{8}$

5 $p = \dfrac{5}{200} = \dfrac{1}{40}$ 6 $p = \dfrac{5}{9}$

7 (a) $p = \dfrac{1}{7}$ (Since 7 different boys could be in that seat and only one of them gives the looked-for result.)

(b) $p = \dfrac{2}{7}$ (Since there are *two* ways of achieving this result.)

Notice that, in all these cases, we had to know about the number of *equally likely* outcomes before we could calculate the theoretical probability. For instance, if you asked people to *name* (rather than choose at random) an odd number between 1 and 9, would you still expect the five possible numbers to be equally likely? Explain.

In fact, the five possible numbers (1, 3, 5, 7, 9) would not be equally likely. Most people you asked would prefer 3 or 7, and the other odd numbers would be mentioned less often. Indeed, 1 and 9 might be entirely ignored. (If you doubt this, try it on a few unsuspecting friends.)

Similarly, in dealing with the cinema seats problem (7), we had to *assume* that each boy was *equally* likely to be in each of the seven different seats. What kind of extra knowledge about those seven boys as individuals might make you

alter your estimate of the theoretical probability ($\frac{1}{7}$) that a named boy was in a

certain seat?

For instance, the knowledge that one boy had a favourite seat, and was powerful enough to make sure he got it, would lead you to give up the theoretical probability which was based on equally likely outcomes. It would no longer be equally likely that all seats were free for Pete Smith.

The essential thing to remember from this section is that: You can't calculate a theoretical probability unless you know the number of ___?___ ___?___ outcomes.

. . . equally likely

Judging by Experience

So far you've been able to work out the theoretical probability of a 'looked-for' result in a number of probability situations. To do this you need to know *how many* possible outcomes the situation has, and whether they are *equally likely*.

In cases where we aren't clear about the number of possible outcomes and whether they are equally likely, theoretical probability can be misleading. Consider this case.

If we toss *two coins*, there appears to be three possible outcomes:
heads on both coins
       or
tails on both coins
       or
heads on one, tails on the other.

What then is the probability of getting *two heads* if we spin the two coins together?

(a) $p = \frac{1}{2}$?   or (b) $p = \frac{1}{3}$?   or (c) $p = \frac{1}{4}$?

(c) $p = \frac{1}{4}$ is the probability of both coins coming up heads. (Although there are three possible outcomes, they are *not* equally likely. In fact, one of them is twice as likely to happen as either of the other two. There are *two* ways of getting heads on one coin and tails on the other. So 'two heads' is one of *four* equally likely outcomes, rather than of three.)

Once we picture the possible outcomes, this all becomes very clear. We can see that the probability of 'two tails' is also $\frac{1}{4}$.

Coin X   Coin Y

What is the probability of getting a head on one coin and a tail on the other?

$\frac{1}{2}$ is the probability of getting a head on one coin and a tail on the other. (Out of four equally likely outcomes, *two* would give the result 'a head and a tail'.)

Many a gambler has grown rich betting with people who see three equally likely outcomes in a throw of two coins! The moral is: when considering the probability that a situation will turn out a certain way, *don't jump to conclusions* about the number of possible outcomes and whether they are equally likely.

Somewhere in the world at this moment a baby (boy or girl) is being born. What is the probability that it is a girl?

(a) $p = \frac{1}{2}$?   or (b) $p > \frac{1}{2}$?   or (c) $p < \frac{1}{2}$?   or (d) we can't tell?

If you said $p = \frac{1}{2}$ you jumped to the conclusion that boys and girls were equally likely; if you said 'we can't tell' you must have realised that boys and girls are not born with equal frequency and that we can't work out the probabilities until we know the relative frequencies; if you said $p > \frac{1}{2}$ you must have assumed that girls are born more frequently than boys.

In fact, it has been known since the 18th century that there are rather more boys born than girls. (The proportions are somewhat evened-up later on by the higher death-rates for males at all ages.) For instance, there are 1060 male births for every 1000 females—a ratio of about 51% to 49%.

So the probability that a baby being born at this moment is a girl is slightly (*less/more?*) than $\frac{1}{2}$.

... slightly *less* than $\frac{1}{2}$. (Again, the message is clear. Just because there are two possible outcomes, it doesn't mean they are equally possible!)

Now let's think about the problems of Ken Wright who has two girl friends. To visit one he takes a north-bound bus, to visit the other, a south-bound bus. He can never make up his mind which girl to visit, so he always lets 'chance' decide for him. He just goes down to the bus-stops and takes the first bus that comes along, whether it is going north or south.

What is the probability that his next visit will be to Miss South?

(a) $p = \frac{1}{2}$? or (b) $p < \frac{1}{2}$? or (c) $p > \frac{1}{2}$? or (d) we can't tell?

(d) we can't tell (Any other answer would indicate you were jumping to conclusions, unless you have inside information about how the buses run in Ken's town.)

I told you that Ken goes down to the bus-stops and simply takes the north-bound or the south-bound bus, whichever happens to come along first. So what you are trying to estimate is the probability that, on any random visit, the south-bound bus will be the first to arrive.

Suppose I now give you a little extra information: The north-bound and south-bound buses run equally often—every 5 minutes.

Can you now estimate the probability that Ken will be visiting Miss South?

(a) $p = \frac{1}{2}$? or (b) $p \neq \frac{1}{2}$? or (c) we still can't tell?

(c) we still can't tell (To have said $p = \frac{1}{2}$ would have been an understandable mistake, but a mistake nevertheless. Despite the fact that the buses run equally often, Ken is not as likely to catch one as the other.)

Why we still can't tell should become clear once you've seen Ken's local bus timetable. The section shown below should give you all the information you need to decide the probabilities.

North-bound	6.0 am	6.5	6.10	6.15	6.20	etc. throughout the day
South-bound	6.4 am	6.9	6.14	6.19	6.24	

However, you know that Ken does not plan his visits by the timetable. He just goes to catch a bus whenever the fancy takes him. What would you now say is the probability that such a random choice of time would put him on the bus towards Miss South?

(a) $p = \frac{1}{5}$?   or (b) $p = \frac{1}{2}$?   or (c) $p = \frac{4}{5}$?   or (d) we still can't tell?

If your answer was $p = \frac{1}{5}$ or $\frac{1}{2}$, or if you decided we still can't tell, look at this diagram showing the time-interval between the buses:

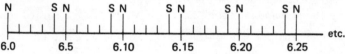

Notice that the interval between a north-bound bus and the next south-bound bus is much *bigger* than the interval between a south-bound bus and the next north-bound. Ask yourself this key question:

If Ken goes to the bus-stop at random times, is he just as likely to arrive there during a time-interval when the next bus along is north-bound as during an interval when the next bus is going south?

So what is the probability he will catch a south-bound bus?

(c) p(that Ken will go south) $= \frac{4}{5}$. (Ken himself reasoned that since the buses run equally often he should have been seeing each girl equally often. In fact, he found he was visiting Miss South about 80% of the time, and Miss North only about 20%. This, as he realised when he gave some thought to the timetable, was because he was four times as likely to arrive at the bus-stops during a time-interval when the next bus along would be the south-bound.

So there were *two* possible outcomes of any one decision to visit a girl friend—Ken either got to see Miss North or he got to see Miss South. Were these two outcomes equally likely?

No, the two outcomes were not equally likely. (One was four times as likely as the other.)

In the last few examples we've seen how easy it is to miscalculate the theoretical probability of a result when we can't be sure about whether possible outcomes are equally likely. Fortunately, there's another method of estimating a basic probability—a method based, not on theory, but on practical experience.

Practical Probability

By this method we look at what actually *has* happened, rather than at what 'ought' to happen. We estimate the probability of a result by counting the number of times it has already occurred in a certain number of 'trials'.

$$\text{Practical probability} = \frac{\text{No. of times result has occurred}}{\text{No. of trials}}$$

Make a *note* of this formula.

The fact that Ken found himself seeing Miss South 33 times in the course of 40 'trials' (visits) led him to realise that the practical probability of seeing her on any future trial was $\frac{33}{40}$ (or about $\frac{4}{5}$). (Provided, of course, that the bus company doesn't change its timetable!)

Similarly, if 514 out of every 1000 babies born are boys, we might say that the practical probability of a randomly-chosen birth being male is __?__ .

$\frac{514}{1000}$ or 0.514 (that is, out of 1000 trials, 514 give the looked-for result.)

So estimates of practical probability are based on *relative frequency* (rather than on equally likely outcomes).

Think about this one.

If you toss a thumbtack onto a table, it can land with its point either 'up' or 'down'.

UP   DOWN

Suppose I try to persuade you that the probability of 'UP' is $\frac{1}{2}$. What would be the most sensible thing for you to do?

  (a)   Agree, because there are two possible ways the tack could fall?

or (b) Disagree, because the tack is not symmetrical like a coin?

or (c) Ask to see the results of throwing a thumbtack a large number of times?

(c) Ask to see the results of throwing the tack a large number of times. (We have no previous experience with thumbtacks that tells us whether 'up' and 'down' are equally likely, so we can't use theoretical probability.)

Let's say we've tossed the thumbtack 100 times, and got 'UP' on 65 occasions, We would now have to estimate the probability of 'UP' on future tosses as being ___?___.

$\frac{65}{100}$ or 0.65 (This is the relative frequency we've recorded in practice.)

Really, I haven't the slighest idea whether this is anything like the 'true' probability. If you have a thumbtack handy, and a few minutes to spare, you might like to make your own estimate based on an actual experiment. (If you have some friends, each of whom is willing to toss a thumbtack, say 100 times, you may find it interesting to compare the relative frequencies recorded by each separate person, and then *pool* the separate results to compare the relative frequency of 'UP' over *all* the tosses taken together.)

Knowledge of what has happened in the past can thus help indicate what might happen in the future. To say that you have an $\frac{8}{10}$ chance of succeeding in a college course does not mean there are eight equally likely ways the course can turn out, and all but two will give you success. Nor does it mean that if you took the course ten times you would succeed eight times.

It simply means that, out of ten people like you who take the course, ___?___ are successful.

eight

What is the probability of surviving a heart transplant operation, say for one year or more? Again, we have no way of working out a theoretical probability. There are two outcomes, certainly: you die or you recover. But there is no reason to suppose they are equally likely. So we must rely on a practical estimate of probability. As Aristotle put it, many years ago: 'The probable is what usually happens'.

Let us suppose 20 people have so far had transplants and 18 of them have died. Based on this, what is the practical probability of *surviving* such an operation?

$p = \frac{1}{10}$ (out of 20 'trials' the looked-for result happened only twice—a practical probability of $\frac{2}{20} = \frac{1}{10}$.)

Practical estimates of probability will vary to some extent with the *number of trials* being considered. (If you did the thumbtack experiment with friends, you'll perhaps have noticed this already.) In the short term, chance factors may bias the results in one direction, but, in a longer run, they get *balanced out* by chance factors leaning in the other direction. For instance, while 18 of the last 20 heart transplant patients failed to survive the operation, perhaps as few as 12 from the next 20 patients will die—even without any improvements in the skill of the surgeons.

Suppose only 12 out of the next group of 20 do die. What will then be our practical estimate of the probability of survival?

(a) $p = \frac{1}{4}$? or (b) $p = \frac{2}{5}$? or (c) $p = \frac{3}{4}$?

$p = \frac{1}{4}$ (We have now seen a total of 2 + 8 survivals in the course of 20 + 20

'trials'. So $p = \frac{2 + 8}{20 + 20} = \frac{10}{40} = \frac{1}{4}$.)

Even so, with such small numbers of trials, chance factors could easily be distorting our estimates of probability. Unfortunately, we have no way of making more reliable estimates on the evidence available to date. To improve our estimates, further practical information is needed.

The effects of chance fluctuation can easily be seen in a coin-tossing experiment. The theoretical probability of 'heads' is $\frac{1}{2}$. Just what does this mean in practice?

Suppose you throw a coin and get 'tails'. What is the probability you will get 'heads' if you throw the coin a second time?

(a) $p < \frac{1}{2}$? or (b) $p = \frac{1}{2}$? or (c) $p > \frac{1}{2}$?

(b) $p = \frac{1}{2}$ (The probability on the second throw is exactly the same as on the

first. The coin has no memory of the last result it gave; and it certainly has no urge to even things up by giving you the other result next time!)

Since each face of the coin stays equally likely on each throw, it is quite usual to get *runs*—several heads or tails in a row. In five throws, for instance, you might get a run of four heads. But this alone should not make you revise your estimate to p(heads) $= \frac{4}{5}$. For in your next five throws you might equally well have a run of four tails that would balance out the heads.

Look below at the results of tossing a coin 1000 times:

	Heads	Tails
After 10 throws, the number of heads and tails were:	6	4
After 1000 throws, the number of heads and tails were:	480	520

After 10 throws, two more heads than tails had been thrown. After 1000 throws, however, there was a difference of 40 between the numbers of heads and tails that had been thrown.

But over which set of throws (the first 10 or the complete 1000) was the practical probability closer to the theoretical probability of p(heads) $= 0.5$?

Despite the gap between heads and tails growing larger as the number of throws increased, the practical probability was closer to the theoretical probability at the end of the 1000 throws:

	Heads	Tails	gap	practical probability of heads
10 throws	6	4	2	0.6
1000 throws	480	520	40	0.48

During the Second World War, a Danish prisoner-of-war passed the time by throwing a coin 10 000 times. He got 5067 heads. This gives a practical probability of 0.5067, which is even closer than we got with 1000 throws to the theoretical probability of 0.5

So, the greater the number of trials, the (*more/less?*) accurate will be your estimate of probability.

---

more (If you have any doubts, try throwing some coins or dice, cumulating the results after each set of, say, 10 throws. If you have friends to work with you, you will be able to amass large amounts of data quite quickly.)

The larger the sample, the more difficult it is for chance fluctuations to distort the relative frequencies within it. And the more likely you are to get the *same* relative frequencies in other such samples.

Below are the results of two coin-tossing experiments—each with a *different* coin. If I told you that one of these coins was biased towards heads, which one would you say I was talking about?

(a) coin X, which gave 7 heads in 10 throws?
or (b) coin Y which gave 700 heads in 1000 throws?
or (c) would you say both coins are clearly biased?

(b) coin Y, which gave 700 heads in 1000 throws, is the biased coin. (Seven heads in ten throws would not be *too* remarkable. You would still feel it was just a fluke, that the probability of heads on the next throw was still $\frac{1}{2}$, and that in the long run the relative frequencies would tend to even out. But 700 heads in 1000 throws is rather different. The frequencies have had plenty of time to even up, yet heads are suspiciously far into the lead.)

What would you reckon to be the probability of getting heads on your next throw with coin Y?

(a) $p < \frac{1}{2}$?   or (b) $p = \frac{1}{2}$?   or (c) $p > \frac{1}{2}$?

(c) $p > \frac{1}{2}$ (A coin giving 700 heads in 1000 throws is almost certainly biased towards heads, and the probability of getting another head on your next throw is surely closer to 0.7 than to 0.5.)

So how do the theoretical and practical estimates of probability compare?

Theoretical probability $= \dfrac{\text{No. of outcomes giving looked-for result}}{\text{Total no. of equally likely outcomes}}$

Theoretical probability gives an exact figure but is based on an *assumption* about equally likely outcomes.

Practical probability $= \dfrac{\text{No. of times result has occurred}}{\text{No. of trials}}$

Practical probability gives an estimate whose *accuracy varies* with the number of trials.

Make sure you have a *copy* of these formulae.

Why do you think it is, then, in spite of this variable accuracy, that the bulk of real-life probabilities are settled by the *practical* method? (For example, the probability that an 18-year old driver of a sports car will have a serious crash this year.)

The reason is that it is very difficult, if not impossible, to find equally likely outcomes in real-life situations. (For example, what equally likely outcomes can we possibly see in the case of the 18-year old and his sports car?)

Gambling, Probability and Belief

We have come to the conclusion that equally likely outcomes are not very helpful in deciding real-life probabilities. In real-life situations we usually work with data about relative frequency. Thus we establish practical probability estimates that can be used to predict the likelihood of future events.

Take insurance companies, for instance. Before they can decide how much premium to charge you for a given amount of insurance coverage, they must look carefully at their figures for recent years to see how frequently they've had to pay out on the kind of insurance policy you want. They will use this past experience to estimate the probability that they will have to pay out again in future. The bigger the probability, the bigger the premium they'll charge you.

Insurance companies charge lower automobile insurance premiums to drivers over the age of 30. This suggests that the probability of such drivers having accidents had been found to be (*higher/lower?*) than that of younger drivers.

lower (Older drivers were found to have a lower accident-rate than younger ones. So the probability of the insurance companies having to pay out to repair the cars of older drivers is smaller, and it may safely charge them a smaller premium for the same amount of 'coverage'.)

Insurance companies base their success on the same 'law of large numbers' that makes it more probable you'd get about 50% heads in 10 000 tosses of a coin than you would in 10 tosses. A company can be confident that, say, 25% of all the sports cars it insures are going to crash and claim money from it, even though it has no more idea than you have of whether one *particular* sports car is going to be among them.

So, while the driver never knows quite what he will get out of the bargain, the insurance company does know what it will gain. As someone once pointed out:

Insurance is a (*gamble/certainty?*) for the person buying it, but a (*gamble/certainty?*) for the company that sells him it!

Insurance is a *gamble* for the person buying it, but a *certainty* for the company that sells him it. In theory, at least, the premiums taken in from all the people who do not make claims should more than cover the costs paid out to people who do. This works best, of course, when the company knows the maximum amount that would have to be paid out in the event of a claim, as happens with fire insurance or medical insurance. Many companies have complained that they lose money on the automobile insurance side of their business when inflation increases the cost of repairs far faster than they can increase premiums.

Talking of gambling, it's interesting to note that the study of probability first got under way when an 18th century gambler began asking questions about the dice that were losing him so much money. And although probability theory has since proved its value in many fields from industrial production to medical research, courses on statistics still refer frequently to the tools of the gambler (dice, coins, playing cards, etc) simply because they give such a clear picture of theoretical probabilities.

While gamblers do need some idea of probability, both theoretical and practical, many of their decisions are based not on calculation but on 'hunch'. Perhaps we are all gamblers in this respect. Consider the following, for instance:

(i)   How would you set about estimating the probability that:
      Seattle Slew will win the Kentucky Derby?
      Britain will stay in the European Economic Community?
      Earth will be invaded by creatures from outer space?
      A current Supreme Court case will be decided in favor of the plaintiff?
      Income tax will be increased next year?

(ii) Would different people all make the same estimates as you?

(i) In none of these cases could you estimate probability by counting equally likely outcomes. But nor could you use the idea of relative frequency over a series of repeated trials. Each event would have been *unique*: (these particular horses only run once in the Kentucky Derby: we can't compare the number of times the creatures from outer space have invaded us with a number of times they haven't; and so on.) And previous experience of Supreme Court Cases and tax changes will not help us calculate a probability: we have to guess.

(ii) Surely different people would guess differently about the probability of each event according to their information, beliefs, hopes, fears, and general cast of mind. One 'expert' might say 'highly likely' while another might say 'out of the question'. This sort of contradictory opinion-mongering is rife at election-times!

Now whether the 'guess' comes from a tipster 'playing a hunch' or from a world-famous expert stating a considered opinion, what we are getting is not a calculation about probability but a statement about *the strength of his personal belief*. You will hope that he has considered any *evidence* that exists (e.g. opinion polls in the case of an election forecast) and weighed up all the factors that might make for one outcome rather than another. In the end, however, whether or not you can rely on someone else's estimate of this kind of probability depends on how much *faith* you can put in his belief.

Perhaps it is true to say that even when the probability is a calculated one, it takes some degree of belief to *act* upon it. Can you be sure that the theory from which you deduced a theoretical probability really applies to the situation you are dealing with? (e.g. is your opponent playing with loaded dice?) Or can you be sure that the conditions are exactly the same as during the trials that gave you a practical probability? (e.g. perhaps thumbtacks land differently if dropped from a greater height?)

Suppose you are a surgeon and a patient comes to ask you about a particular operation. You know that 98 out of every 100 people who have the operation make a perfect recovery. Which of the following comments might you make to your patient, and why?

   (a) 'You're practically certain to pull through'?

or (b) 'I shouldn't advise *you* to risk it'?

or (c) I might make either comment?

(c) I might make either comment. (You would surely apply the practical probability either to encourage or to dissuade the patient according to whether you believed him typical of the 98% successes or the 2% failures.)

What Are the Odds?

Gamblers and guess-makers often express their hunches about probability not as fractions but in terms of *odds*. They speak of the *odds against* something happening, or the *odds in favour*. Odds of *2 to 1 against* 'Seattle Slew' winning the Kentucky Derby would suggest that he is thought to have two chances of losing for every chance of winning. Out of three chances, two are unfavourable. In short, the probability of losing is 'guesstimated' to be $\frac{2}{3}$.

If the odds against 'Secretariat' are 5 to 3, what is thought to be the probability the horse will lose?

(a) $q = \frac{3}{5}$?   (b) $q = \frac{5}{8}$?   or (c) $q = \frac{3}{8}$?   or (d) $q = \frac{2}{8}$?

(b) $q = \frac{5}{8}$ (If the odds against are 5 to 3, the horse is thought to have 5 chances of losing as against 3 chances of winning. So, out of $5 + 3 = 8$ chances, 5 will result in a loss: $q = \frac{5}{8}$.)

To state the odds *in favour* of an event, of course, we write the figures the other way round. The odds in favour of 'Secretariat' *winning* are 3 to 5. (3 chances of winning compared with 5 of losing.) This is sometimes called '3 to 5 on', or '3 to 5 for'. (In probability language, $p = \frac{3}{8}$.)

To end this chapter, here are a few experiments you can do for yourself, so you can get an even better idea of how probability (both theoretical and practical) works out in practice. Carry out as many of these experiments as you have time for. In most cases it would be helpful if you could compare notes with colleagues doing the same experiment as yourself, thus giving everyone a lot more data to work with.

Experiments in Probability

1   Two dice, thrown together, can produce totals between 2 and 12. But not all the totals are equally likely. Throw a pair of dice 100 times, and count the number of times each possible total appears. What are the practical probabilities of the various totals? Can you work out the theoretical probabilities also, and compare?

2   Ask 30 or more people to draw a line which they estimate to be 10 inches long. Note whether each person under-estimates or over-estimates. Use the results to calculate a practical probability for each of those outcomes, and check it by testing another set of people in the same way.

3   Roll a 6-sided pencil down an incline (say a sloping sheet of cardboard) 50 times. Note the spot where the pencil stops after each roll, and measure

the distance it has travelled. Record these distances in a table and/or a block diagram (histogram). (The results should approximate to a 'normal' distribution.)

4 Take a matchstick (or toothpick, cocktail stick, needle, etc.) and rule off parallel lines on a sheet of paper in such a way that the distance between the lines is *twice* the length of the matchstick. Toss the matchstick onto the ruled surface, without deliberate aim, at least 200 times. Record the total number of tosses and the number of times the matchstick touches or crosses one of the lines. What then do you estimate as the practical probability of touching or crossing? You should find this is close to $p = \dfrac{1}{\pi}$ (you'll remember $\pi =$ 3.142... from the circle formula $C = 2\pi r$). Toss the matchstick 100 times more and see if this brings you even closer to $p = \dfrac{1}{\pi}$.

5 If you are working in a school or college, select ten students at random from a class and ask each one whether or not he has had his appendix removed. Use this information to estimate how many students in the class as a whole will have had their appendices removed. Then check your estimate by asking the remaining students.

6 There is a traditional game, sometimes called 'Scissors, Paper, Rock', in which the two players, at an agreed signal, each brings out a hand from behind his or her back showing one of three possible 'patterns'. Each player either holds out two fingers (scissors) or holds their hand out flat (paper) or makes a fist (rock). Since scissors cut paper, scissors win over paper; since paper can wrap up a rock, paper wins over rock; since rock blunts scissors, rock wins over scissors. If both players show the same pattern, that 'round' is drawn. Play the game with a colleague—about 30 rounds should be enough. Call yourself Player 1 and record the number of times Player 1 wins, the number of times Player 2 wins, and the number of times the round is drawn. What, then, is the practical probability for each of these outcomes? (Could you get the same sort of probability situation using two dice instead of hands?)

7 Throw two coins together 50 times. Keep a tally of the outcomes. Do the outcomes '2 heads', '2 tails', and '1 head/1 tail' appear to be equally likely? Explain.

8 Toss a coin 100 times. After each set of ten throws, record the number of heads during that set, and also the total number so far. Calculate the practical probability as suggested by the latest set of throws and compare it with the more reliable (?) estimate you would make in the light of all the trials so far. How does the latter estimate change as the number of trials increases?

Carry out as many of these experiments as possible. Discuss your results with teachers and colleagues, if you can. If you have time, design some probability experiments of your own, discuss them with teachers and colleagues, and get their help in carrying them out.

2

Combined Probabilities

Total Probability

Most of our thinking so far has been about *simple* probabilities—where a single result can happen or fail to happen. We must now go on to consider the probabilities that arise when we are interested in two or more results at once. We will find, for instance, that the probability of getting an odd number OR a '6' in one throw of a die is very different from the probability of getting an odd number AND a '6' in two throws.

Suppose that to win a certain game you must throw a '4' with a die. What is your theoretical probability of doing so?

Theoretical probability $= \dfrac{\text{No. of outcomes giving looked-for result}}{\text{Total no. of equally likely outcomes}} = \dfrac{1}{6}$.

Since the die could equally well turn up in any of six ways, and only one of them is of use to you, you have a one in six chance of winning. Similarly, if it was, for example, a '6' you needed to win, your chances would be exactly the same.

But let's suppose it doesn't matter whether you throw a '4' OR a '6'. Suppose either of these will win you the game. Your probability of winning would then be: (a) greater? or (b) smaller? or (c) exactly the same?

(a) greater (If either of *two* possible results will win the game for you, then you clearly have more chance of winning than you would if only one of those results would be of use.)

When we are trying to calculate the probability that *either* one *or* the other of two (or more) results will occur, we call it the *total* probability.

So what is the total probability of throwing *either* a 4 *or* a 6?

$p = \dfrac{1}{3}$ is the probability of throwing either a 4 or a 6. That is:

$\dfrac{\text{No. of outcomes giving looked-for result (4 or 6)}}{\text{Total no. of equally likely outcomes}} = \dfrac{2}{6} = \dfrac{1}{3}$.

Notice that the total probability of either one or the other results (throwing a 4 or throwing a 6) is the SUM of their separate probabilities:

$$\frac{1}{6} + \frac{1}{6} = \frac{2}{6}$$

What is the probability of throwing an even number—that is, either a 2 or a 4 or a 6?

$p(2 \text{ or } 4 \text{ or } 6) = \dfrac{1}{2}$ That is:

$$\frac{\text{No. of outcomes giving 2 or 4 or 6}}{\text{Total no. of equally likely outcomes}} = \frac{3}{6} = \frac{1}{2}$$

Again, notice that the total probability of getting one or other of these results is equal to the SUM of their separate probabilities:

$$\left. \begin{array}{l} p(2) = \dfrac{1}{6} \\[2mm] p(4) = \dfrac{1}{6} \\[2mm] p(6) = \dfrac{1}{6} \end{array} \right\} \quad p(2 \text{ or } 4 \text{ or } 6) = \frac{1}{6} + \frac{1}{6} + \frac{1}{6} = \frac{3}{6} = \frac{1}{2}$$

This is because the results are *mutually exclusive*. That is, if any one of these results happens it automatically prevents the others from happening at the same time. If the number thrown is a 4, it can't at the same time be a 2 or a 6. With these three results, each excludes the others.

Here is another example of mutually exclusive results. Suppose I draw one card from a pack of playing cards. I may draw a King or an Ace. But clearly the card I have drawn cannot be both King and Ace. So the two possibilities are mutually exclusive—whichever one I get, it prevents the other from happening.

But you might point out that the result of my drawing one card might be *a King or a red card*. These two possibilities are NOT mutually exclusive. If I draw a King it might *at the same time* be a red card (King of hearts or diamonds), and vice versa. Similarly, 'a Jack or a face card' would *not* be mutually exclusive, but 'a face card or a numbered card' *would* mutually exclude one another. For practice, decide whether each of the pairs of possible results below are mutually exclusive or not.

Drawing:
(i)   a diamond or a black card
(ii) a 10 or a Jack
(iii) a face card or a Queen
(iv) a spade or a black card
(v)   a card higher than 5 or a card lower than 9.

(i) and (ii) are mutually exclusive; the others are not.

Below, you'll see I've described several possible results of throwing a die. In which pair are the two results *mutually exclusive*?

(a) Throwing an odd number OR Throwing a 3?
or (b) Throwing a multiple of 2 OR Throwing a multiple of 3?
or (c) Throwing a 5 OR Throwing an even number?
or (d) Throwing a 6 OR Throwing a number greater than 4?

(c) Throwing a 5 OR Throwing an even number are mutually exclusive results.
(If the number thrown is even (2, 4, or 6) it can't at the same time be a 5. Since the result 'throwing an even number' excludes the possibility of 'throwing a 5', and vice versa, the two results are said to be mutually exclusive.)

This brings us to our first rule for combining probabilities:

Addition Rule

> If two or more results are MUTUALLY EXCLUSIVE, the TOTAL probability that one OR the other will happen is the SUM of their separate probabilities.

Make a *note* of this rule.

Use the addition rule to find the total probability of throwing either a 5 or an even number.

$p(5 \text{ or even}) = \frac{2}{3}$. Since the results are mutually exclusive, the total probability of one or the other happening is the sum of their separate probabilities:

$$\left.\begin{array}{l} p(5) \quad = \frac{1}{6} \\[2mm] p(\text{even}) = \frac{1}{2} \end{array}\right\} p(5 \text{ or even}) = \frac{1}{6} + \frac{1}{2} = \frac{1+3}{6} = \frac{4}{6} = \frac{2}{3}$$

You can check the result we've got from the addition rule by using our original formula for theoretical probability:

$$\frac{\text{No. of outcomes giving looked-for result (5 or even)}}{\text{Total no. of equally likely outcomes}} = \frac{4}{6}$$

In tossing a coin, heads and tails are mutually exclusive results. The coin can't land both ways up at once.

What is the probability of getting one result or the other?

$p(\text{head } or \text{ tail}) = 1.$
That is:

$$\left.\begin{array}{l} p(\text{head}) = \frac{1}{2} \\[2mm] p(\text{tail}) \quad = \frac{1}{2} \end{array}\right\} p(\text{head or tail}) = \frac{1}{2} + \frac{1}{2} = 1$$

That is why the coin-tosser who says 'Heads I win, tails you lose' is onto such a good thing. He is certain of getting *either* one *or* the other.

So, to find the total probability of getting either one or the other of several mutually exclusive results, we add their individual probabilities of happening. Not surprisingly, this kind of combined probability is often called EITHER ... OR probability.

What is the total probability of selecting *either* a black *or* a red *or* a white ball from a bag containing 2 black, 3 red, 5 white, 20 green?

$$p(\text{B or R or W}) = \frac{1}{3}$$

There are 30 balls: 2 black, 3 red, 5 white, and 20 green. Whatever colour ball we choose it excludes the possibility of having chosen some other colour—they are mutually exclusive.

$$\left. \begin{aligned} p(\text{Black}) &= \frac{2}{30} \\ p(\text{Red}) &= \frac{3}{30} \\ p(\text{White}) &= \frac{5}{30} \end{aligned} \right\} p(\text{B or R or W}) = \frac{2}{30} + \frac{3}{30} + \frac{5}{30} = \frac{10}{30} = \frac{1}{3}$$

Now apply the addition rule to these problems:

(i)   What is the probability of drawing either a face card or a 10 from a pack of 52 playing cards?

(ii) What is your chance of throwing a number less than 3 or a number greater than 6 in one throw of a die?

(iii) If there are eight chocolates left in a box, including two with marzipan centres and three others with cream centres, what is the probability that a random draw would give you a cream or marzipan centre?

(i) $\left. \begin{aligned} p(\text{face}) &= \frac{12}{52} \\ p(10) &= \frac{4}{52} \end{aligned} \right\} p(\text{face or } 10) = \frac{12}{52} + \frac{4}{52} = \frac{16}{52} = \frac{4}{13}$

(ii) $\left. \begin{aligned} p(<3) &= \frac{2}{6} \\ p(>6) &= \frac{0}{6} \end{aligned} \right\} p(<3 \text{ or } >6) = \frac{2}{6} + \frac{0}{6} = \frac{2}{6} = \frac{1}{3}$

(iii) $\left. \begin{aligned} p(\text{M}) &= \frac{2}{8} \\ p(\text{C}) &= \frac{3}{8} \end{aligned} \right\} p(\text{M or C}) = \frac{2}{8} + \frac{3}{8} = \frac{5}{8}$

By the way, could you have solved those last three problems by using our original formula for theoretical probability?

That is: $\dfrac{\text{No. of outcomes giving looked-for result}}{\text{Total no. of equally likely outcomes}}$?

Yes, you could have solved those problems by using our original formula for theoretical probability. For example, in problem (iii) there were 2 marzipan and 3 cream centres, giving 5 outcomes that would provide the looked-for result, compared with 8 possible outcomes (all the chocolates) altogether: $\dfrac{5}{8}$ as before. (If you doubted whether the formula could be used, try it on the other two problems yourself.)

So the addition rule is another way of describing something you can already do. Why, then, is it worthwhile your knowing about it?

Two reasons:

1 In later problems you'll find it more useful than the formula.
2 It introduces you to the idea of mutually exclusive results and will help you see the difference between total probability and the other way of combining probabilities that we'll be talking about in the next section.

So, according to the addition rule:

When two or more results are __?__ __?__ we can find the total probability of one or the other happening by __?__ ing their separate probabilities.

---

mutually exclusive ... add

Never try to apply the addition rule to find the combined 'either ... or' probability unless you are *quite sure* that the results really are mutually exclusive.

For instance, three girls are being interviewed for entrance to college. Each has a different combination of honors subjects:

	honors subjects
Sue	Sociology, History
Lena	Economics, History
Clare	Economics, Statistics

What is the probability that the first girl to be interviewed will have honors Economics or History?

(a) $p = \dfrac{1}{3}$? or (b) $p = \dfrac{2}{3}$? or (c) $p = \dfrac{3}{3} = 1$? or (d) $p = \dfrac{4}{3}$?

$p = \dfrac{3}{3} = 1$ (As you can see from the list of honors, *every* candidate has either Economics or History, or both. So whoever is interviewed first is *certain* to have one or the other.)

If you tried to apply the addition rule to that problem, you would reason that two of the three girls have Economics, so the probability of the first candidate having this subject is $\frac{2}{3}$. Similarly, two out of the three have History, so the probability here also is $\frac{2}{3}$. If you added these probabilities you would get $\frac{2}{3} + \frac{2}{3} = \frac{4}{3}$ which is clearly absurd since you can't have a probability greater than 1.

And why would the addition rule give an absurd answer in this case? Because the possible results whose probabilities are being added (*are/are not?*) mutually exclusive.

---

are not (Lena has both Economics and History, so the addition would include her *twice*. The real probability is the sum of the two separate probabilities *minus* the probability that the chosen candidate has both subjects: $\frac{2}{3} + \frac{2}{3} - \frac{1}{3} = 1$.)

Where possible, use the addition rule to find the probability of:
  (i)   selecting blindfolded a yellow rose or a white rose from a vase containing 3 yellow roses, 2 pink roses, 4 red roses, and 6 white roses.
  (ii) drawing a card numbered 2, 3, 4, or 5 from a pack of 52 playing cards.
  (iii) choosing at random a card that is a multiple of 2 or 5 from a set of twenty cards marked 1 to 20. (Draw up a table.)

(i) $\left. \begin{array}{l} p(Y) = \dfrac{3}{15} \\[2ex] p(W) = \dfrac{6}{15} \end{array} \right\}$ $p(Y \text{ or } W) = \dfrac{3}{15} + \dfrac{6}{15} = \dfrac{9}{15} = \dfrac{3}{5}$

(ii) $\left. \begin{array}{l} p(2) = \dfrac{4}{52} \\[2ex] p(3) = \dfrac{4}{52} \\[2ex] p(4) = \dfrac{4}{52} \\[2ex] p(5) = \dfrac{4}{52} \end{array} \right\}$ $p(2 \text{ or } 3 \text{ or } 4 \text{ or } 5) = \dfrac{4 + 4 + 4 + 4}{52} = \dfrac{16}{52} = \dfrac{4}{13}$

(iii) The correct answer is NOT $\dfrac{7}{10}$.

It's *no good* saying there are ten multiples of 2, and four multiples of 5 (giving probabilities of $\dfrac{10}{20}$ and $\dfrac{4}{20}$), and if we add them together we get $\dfrac{14}{20}$ as the total probability of choosing one or the other.

As you can see, if we lay out the twenty numbers, only twelve of them are multiples of 2 or 5; so the probability of our choosing one of them is $\frac{12}{20} = \frac{3}{5}$

1 ② 3 ④ ⑤ ⑥ 7 ⑧ 9 ⑩
11 ⑫ 13 ⑭ ⑮ ⑯ 17 ⑱ 19 ⑳

The reason we can't apply the addition rule to this problem is that the two looked-for results (choosing a multiple of 2, and choosing a multiple of 5) are NOT __?__ __?__ results.

... NOT *mutually exclusive* (Two of the numbers we might choose are multiples of both 2 and of 5 at the same time—10 and 20; so the two results would not exclude one another.)

That's enough for the moment about total probability. But we'll take up the idea again later, after the next section in which we look at another way of combining probabilities.

Joint Probability

In the last section we considered how to combine separate probabilities in order to arrive at *total* probability—the probability that *one or other* of several possible results would occur. We noticed that this kind of probability is sometimes called 'either ... or' probability.

In this section we'll look at another kind of combined probability—the probability that two or more results will happen together. This is sometimes called 'both ... and' probability or, more properly, *joint* probability.

Suppose you are playing two coin-tossing games. Here is what you would have to do to win each game:

Game 1   Throw *either* a head *or* a tail in *one* toss of the coin.
Game 2   Throw a head *and* a tail (in that order) in *two* tosses.
Which game would you be more likely to win?

Game 1 (You'd be certain to throw either a head or a tail, and so win Game 1. But in two tosses of a coin you are *not* certain to throw both a head and a tail. You might have two heads or two tails; even if you had one of each they might be in the wrong order.)

Now try your hand at Games 3 and 4:

Game 3   Throw *both* 4 *and* 6 (in that order) in *two* tosses of a die.
Game 4   Throw *either* 4 *or* 6 in *one* toss of a die.
Again, which game would you be more likely to win?

Game 4 (You would be more likely to throw either a 4 or a 6 in one toss than both a 4 and a 6 in two tosses. If you're not quite sure why, you soon will be.)

In each game, then, we've considered two possible results:
Game 1   *Either* heads *or* tails   Game 2   *Both* heads *and* tails
Game 3   *Either* 4 *or* 6   Game 4   *Both* 4 *and* 6

So, over both pairs of games, the probability of getting both one result *and* the other has been (*higher/lower?*) than the probability of getting either one result *or* the other.

... lower

Let's work out why the joint probability should be lower by examining our heads/tails game. First, what is the total probability of throwing either a head or a tail in one toss? Clearly, since the two outcomes are mutually exclusive, their total probability is the sum of their separate probabilities: $\frac{1}{2} + \frac{1}{2} = 1$.

But what is the probability of both a head *and* a tail (in that order) in two tosses of a coin? Before we work it out, are these two results mutually exclusive?

No. (Getting heads on the first throw does not stop you getting tails on the second—or vice versa, obviously.)

So we can't add the separate probabilities to get the combined probability. Let's look at it from another angle: how many equally likely outcomes are there from throwing a coin twice? And how many of these outcomes will give you the looked-for result—a head on the first throw followed by a tail on the second? Once you know these numbers you can use the probability formula.

Let's make a start by listing the possible outcomes. For instance, the first throw could land heads (H) followed by heads on the second throw also.
Complete the table for the other possible joint outcomes.

1st throw	2nd throw
H	H

Here is the table showing all the possible joint outcomes from throwing a coin twice:

	1st throw	2nd throw
Head followed by Head	H	H
Head followed by Tail	H	T
Tail followed by Head	T	H
Tail followed by Tail	T	T

So:
  (i)   How many possible joint outcomes are there from throwing a coin twice?
  (ii) Are all these joint outcomes equally likely?
  (iii) How many of these outcomes give head and tail in that order?
  (iv) Then what is the probability of getting both a head and a tail (in that order) in two throws of a coin?

(i)   There are 4 possible outcomes.
(ii) They are all equally likely.
(iii) Only 1 of them gives head followed by tail.
(iv) So the probability of getting head and then tail is:

$$\frac{\text{No. of outcomes giving looked-for result (H then T)}}{\text{Total no. of equally likely outcomes}} = \frac{1}{4}$$

Here we have combined the two separate probabilities:

$$p(\text{head}) = \frac{1}{2} \quad \text{and} \quad p(\text{tail}) = \frac{1}{2}$$

to get a JOINT probability of $\frac{1}{4}$. (To get the combined result we were looking for, Head followed by Tail, two equally likely outcomes had to be joined together—BOTH had to occur.)

What is the joint probability of getting *two heads* in *two* throws?

Looking back at the table, we can see that, out of 4 equally likely outcomes from two throws, only 1 gives H followed by H. So the joint probability is again $\frac{1}{4}$.

Just one point before we go on: we worked out the table to show all the equally likely joint outcomes that are possible if you throw a coin twice.

Suppose we re-label the columns as shown alongside. Can we now use the same results to represent the possible equally likely joint outcomes from throwing *two coins* together?

1st coin	2nd coin
H	H
H	T
T	H
T	T

Yes. (Two coins thrown together give just the same possible equally likely outcomes, and just the same joint probabilities, as two throws of a single coin.)

Consider another couple of games and what you would have to do to win:
Game 5   Throw two coins and get a tail on both.
Game 6   Throw one coin twice and get a tail each time.
    What is your probability of winning in each game?

In each game your probability of winning is $\frac{1}{4}$. (Whether you throw two coins at once or one coin twice, 'two tails' is one of four possible joint outcomes—HH, HT, TH, and TT.)

Let's look back now at the other pair of games (with dice, this time) that we considered at the beginning of this section:

Either 4 or 6 on one throw of a die

versus

Both 4 AND 6 (in that order) on two throws.

First, what is the total probability of getting either 4 or 6 on one throw?

$\frac{1}{3}$ (Since throwing a 4 or throwing a 6 are mutually exclusive results we add

their separate probabilities, $\frac{1}{6} + \frac{1}{6}$, to get the probability that either one or the

other will happen.)

But when you throw a die *twice* (or two dice together) 4 and 6 are NOT mutually exclusive results. That is, if you throw a 4 first time there is nothing to stop you throwing a 6 second time. So let's work out the chances of this double result happening.

Again we might ask how many equally likely outcomes there are from two throws. Clearly, the first throw could turn up any number of dots from 1 to 6. Is this true of the second throw also?

Yes, the two throws are quite independent—your result on one throw has no effect on the other.

So, the first die could show a 1, with the second coming up 1, or 2, or 3, or 5, or 6. Or the first die could show a 2, with the second coming up 1, or 2, or 3, or 4, and so on.

I have started to write this out in the following table. Each pair of numbers in the body of the table represents one possible sequence of scores from the two dice:

		Score on FIRST DIE				
	1	2	3	4	5	6
1	1,1	2,1				
2	1,2	2,2				
Score on 3	1,3	2,3				
SECOND 4	1,4	2,4				
DIE 5	1,5					
6	1,6					

Please *complete* the table to show all the possible outcomes.

Check your completed table:

		Score on FIRST DIE				
	1	2	3	4	5	6
1	1,1	2,1	3,1	4,1	5,1	6,1
2	1,2	2,2	3,2	4,2	5,2	6,2
3	1,3	2,3	3,3	4,3	5,3	6,3
Score on 4	1,4	2,4	3,4	4,4	5,4	6,4
SECOND 5	1,5	2,5	3,5	4,5	5,5	6,5
DIE 6	1,6	2,6	3,6	4,6	5,6	6,6

So:
1   How many equally likely outcomes are there from throwing two dice (or one die twice)?
2   In how many of these equally likely outcomes is 4 followed by 6?
3   So what is the probability of getting 4 and 6 (in that order) in two throws?

$\frac{1}{36}$ is the probability of a 4 followed by a 6 in two throws of a die. (Out of 36 equally likely outcomes there is only one way you can get 4 followed by 6.)

Clearly, the *joint* probability (both ... and, with two throws) is very much smaller than the total probability (either ... or, with one throw).

Total: Either 4 or 6 in one throw $= \frac{1}{6} + \frac{1}{6} = \frac{1}{3}$.

Joint: 4 and 6 in that order in two throws $= \frac{1}{36}$.

So joint probabilities (like total probabilities) can be calculated by using the formula for equally likely outcomes. Let's try another example.

We've already decided that the probability of throwing two heads in two tosses of a coin is $\frac{1}{4}$. What do you think the probability of throwing three heads in three tosses will be ?

(a) less than $\frac{1}{4}$?   or (b) exactly $\frac{1}{4}$?   or (c) more than $\frac{1}{4}$?

(a) less than $\frac{1}{4}$. (Throwing three heads in three tosses will be even more unlikely than throwing two heads in two tosses.)

Probability Tree

But how much more difficult ? Again we'll draw up a list of equally likely outcomes. The last couple of times we used a table. Another valuable way of keeping your ideas straight, however, is to use a 'probability tree'. Like this:

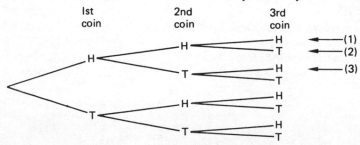

This shows that either H or T on the first throw can be joined with either H or T on the second throw which can be joined with either H or T on the third.

To read the probability tree we start with, say, H in the left-hand column and follow a connecting line across until we have a sequence of three coins. So, the sequence labelled (1) gives HHH. Sequence (2) gives HHT. Sequence (3) gives HTH. And so on. Each of the sequences shown by the connecting lines is an equally likely outcome for a throw of three coins (or one coin thrown three times).

Write out all the equally likely outcomes. How many are there?

There are 8 equally likely outcomes. (By following the lines from column to column we can pick out the eight possible sequences in which the three coins can fall.)

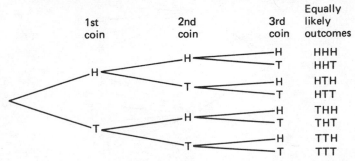

How many of these joint outcomes give three hands in a row? So what is the probability of getting head *and* head *and* head in three tosses?

Only one of the eight equally likely outcomes gives HHH, so the probability of throwing three heads in a row is $\frac{1}{8}$.

Use the probability tree to answer these questions:
What is the probability of getting:
   (i)   Three tails in a row?
   (ii) Two tails and a head (in that order)?
   (iii) Two tails and a head (in *any* order)?

(i) There is only one way of getting TTT, so the probability is $\frac{1}{8}$.

(ii) There is only one way of getting two tails and a head in that order (TTH), so this probability is also $\frac{1}{8}$.

(iii) If we look for the outcomes with two tails and one head in *any* order, we see there are three (HTT, THT, TTH). So the probability is $\frac{3}{8}$. Check that it would be the same for two heads and a tail in any order.

So, we've managed to calculate all these joint probabilities by using the formula for equally likely outcomes. Before we go on to develop a short cut, let's try one last example of this method.

A game consists of tossing a coin and then throwing a die. How many equally likely outcomes does the game have? Draw a probability tree and list the possible outcomes.

Check your probability tree, which should show 12 equally likely outcomes.

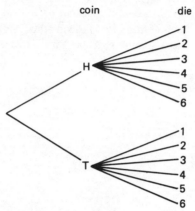

coin   die

(The outcomes are H1, H2, H3, etc. and T1, T2, T3, etc.)

(By the way, if you tried the 'Scissors, Paper, Rock' game on page 25, you may now like to draw a probability tree for the two players to give a theoretical check on your practical estimate or probability.)

Now the probability of getting heads on tossing the coin is $\frac{1}{2}$. And the probability of getting a 3 on throwing the die is $\frac{1}{6}$. But what is the probability of getting *both* heads *and* a 3 in the two throws?

The probability is $\frac{1}{12}$. (Out of 12 equally likely outcomes, only one would give you the looked-for result.)

We've seen how to calculate the probability of two or more results *all* happening by comparing this outcome with the total number of equally likely outcomes. However, it can get a bit tedious working out all the outcomes, so let's look for a short cut.

Check through this summary of examples we've tackled so far:

Probabilities of separate results	JOINT probability of combined result
$p(\text{head}) = \dfrac{1}{2}$ $p(\text{tail}) = \dfrac{1}{2}$	$\dfrac{1}{4} = p(\text{head followed by tail})$
$p(4) = \dfrac{1}{6}$ $p(6) = \dfrac{1}{6}$	$\dfrac{1}{36} = p(4 \text{ followed by } 6)$
$p(\text{head}) = \dfrac{1}{2}$ $p(\text{head}) = \dfrac{1}{2}$ $p(\text{head}) = \dfrac{1}{2}$	$\dfrac{1}{8} = p(\text{head followed by head followed by head})$
$p(\text{head}) = \dfrac{1}{2}$ $p(3) = \dfrac{1}{6}$	$\dfrac{1}{12} = p(\text{head followed by 3})$

What is the arithmetical connection between the size of the two or three separate probabilities and the size of their joint probability?

Did you notice that, in each case, the combined probability was the *product* of the separate, individual probabilities?:

$$p(\text{H followed by T}) = \frac{1}{2} \times \frac{1}{2} = \frac{1}{4}$$

$$p(4 \text{ followed by } 6) = \frac{1}{6} \times \frac{1}{6} = \frac{1}{36}$$

$$p(\text{H and H and H}) = \frac{1}{2} \times \frac{1}{2} \times \frac{1}{2} = \frac{1}{8}$$

$$p(\text{H followed by 3}) = \frac{1}{2} \times \frac{1}{6} = \frac{1}{12}$$

What you have just discovered is called the *multiplication rule*.

Multiplication Rule

There are many ways of putting this rule, but the one below will do best for our purposes in this book:

> Given two or more separate results that are NOT mutually exclusive, the probability of them ALL happening is the PRODUCT of their separate probabilities.

Make a *note* of the rule.

Notice that the rule will *not* apply to mutually exclusive results. As you'll remember, if one such result happens, it prevents the others from happening. So the probability of two or more mutually exclusive results happening is zero (e.g. a coin can't land with both heads and tails uppermost).

Suppose you had to calculate the combined probability of each pair of results below. In which case, (a) or (b) (or both or neither), would you use the *multiplication rule*?

(a) Drawing a King or a Queen from a pack of cards?

or (b) Drawing a King from one pack of cards and a Queen from a second pack?

Only in case (b) would you use the multiplication rule. Here you are trying to calculate a joint probability and the two separate results do not exclude one another. (In case (a) however, it was a mater of choosing one card *or* the other—mutually exclusive possibilities, for which the *addition* rule would be used.)

Very well, imagine you have those two packs of 52 cards each. Since there are 4 Kings and 4 Queens in each pack, you have a $\frac{1}{13}$ chance of getting a King and a $\frac{1}{13}$ chance of getting a Queen.

What is the probability of drawing a King from the first pack and a Queen from the second? (Use the multiplication rule.)

The probability of drawing a King from the first pack and a Queen from the second is $\frac{1}{169}$. (Since $p(K) = \frac{1}{13}$ and $p(Q) = \frac{1}{13}$, and these are quite separate results, we find the probability that both will happen by using the multiplication rule: $\frac{1}{13} \times \frac{1}{13} = \frac{1}{169}$. In fact, not much chance at all. On average, you would only draw King and Queen once in every 169 tries.)

This last probability was a lot easier to calculate using the multiplication rule than if you had tried to work out all the 169 equally likely outcomes. Just picture the amount of space you would need for the probability tree!

Let's try another example. Suppose you are offered three bags, each containing a dozen diamonds. Three of the diamonds in each bag are fakes. You are asked to take one diamond from each bag, but you can have no idea whether you are taking out a real diamond or a fake one. What is the probability that you will choose three *fake* diamonds?

At each attempt your chance of choosing a fake is $\frac{3}{12} = \frac{1}{4}$. Since your three attempts will produce three quite separate results, the joint probability of getting three fakes is given by the multiplication rule: $\frac{1}{4} \times \frac{1}{4} \times \frac{1}{4} = \frac{1}{64}$.

Of course, we could equally well use the multiplication rule to find your chance of getting three *real* diamonds. First of all, if your chance of getting a fake diamond on *one* draw is $\frac{1}{4}$, what is your chance of drawing a real one?

Your chance of drawing a real diamond is $\frac{3}{4}$. (That is, your chance of failing to draw a fake: $p + q = 1$.)

So, if $\frac{3}{4}$ is the probability of drawing a real diamond, what is the joint probability of drawing a real diamond three times in a row—one from each bag?

$p = \frac{27}{64}$. Since $\frac{3}{4}$ of the diamonds in each bag are real, the joint probability of drawing three real diamonds—one from each bag—is $\frac{3}{4} \times \frac{3}{4} \times \frac{3}{4} = \frac{27}{64}$.

So we've calculated two different joint probabilities:

$$p(\text{Fake and Fake and Fake}) = \frac{1}{64}$$

$$p(\text{Real and Real and Real}) = \frac{27}{64}$$

This means that, on average, out of 64 attempts we would select three fake diamonds on one occasion, and three real diamonds on 27 occasions. What would we select on the other 36 occasions?

On the remaining 36 occasions ($1 + 27 + 36 = 64$) we'd select a *mixture* of fake and real diamonds (that is, 2 fake and 1 real, or 1 fake and 2 real).

With that last example of the diamonds we calculated *two* joint probabilities: not only the probability that all three results would happen (choosing three fake stones) but also the probability that they would NOT happen ('failing' three times in a row, and getting three real diamonds instead).

See if you can now apply this to the example before last. You'll remember that the probability of drawing a King from one pack of cards is $\frac{1}{13}$ and that of

drawing a Queen from another pack is also $\frac{1}{13}$. What then is the joint

probability of a *failure* on the first draw being followed by another failure on the second?

The probability of drawing neither a King on the first draw nor a Queen on the

second is $\frac{144}{169}$. Since $p(K) = \frac{1}{13}$ and $p(Q) = \frac{1}{13}$, we can calculate the prob-

ability of failing to get a King on the first draw and of failing to get a Queen on

the second: $p(\text{No K}) = \frac{12}{13}$ and $p(\text{No Q}) = \frac{12}{13}$. Since the two results (failures)

are separate, we need only multiply their probabilities together to find the

joint probability of both failures occurring: $\frac{12}{13} \times \frac{12}{13} = \frac{144}{169}$.

Now let's try picking the winners of six hockey games. Here you are given six games and you are expected to forecast whether each will end in a home win, an away win, or a draw. If we assume the three possible out-

comes are equally likely, you will have a $\frac{1}{3}$ chance of correctly forecasting

the result of a single game.

HOME	AWAY	HOME WIN	AWAY WIN	DRAW
ISLANDERS	MAPLE LEAFS			
FLYERS	RED WINGS			
DEVILS	RANGERS			
CANADIANS	PENGUINS			
NORTH STARS	BRUINS			
OILERS	FLAMES			

If you make one forecast for each game, what is the probability you will forecast all six games correctly?

The probability of all six forecasts being correct is $\frac{1}{729}$, assuming the three

outcomes of any one game are equally likely. Since the six games give six separate outcomes, the joint probability of six correct forecasts is the product of their individual probabilities:

$$\frac{1}{3} \times \frac{1}{3} \times \frac{1}{3} \times \frac{1}{3} \times \frac{1}{3} \times \frac{1}{3} = \frac{1}{729}$$

To be sure of a win on picking six winners, you would need to make 729 different sets of six forecasts.

Try this one.

A factory worker looks after three separate machines. Each has a different probability of breaking down during a shift:

$$\text{Machine A: } \frac{2}{15} \qquad \text{Machine B: } \frac{1}{10} \qquad \text{Machine C: } \frac{3}{11}$$

What is the probability that all three machines will break down during one shift?

The joint probability is $\frac{1}{275}$. To find the probability that all three machines will break down during a shift, we must combine their separate probabilities of beaking down. And, since it is the *joint* probability we are looking for, we'll use the multiplication rule:

$$\frac{2}{15} \times \frac{1}{10} \times \frac{3}{11} = \frac{1}{275}$$

So, on average, they'll all break down together only once in 275 shifts.

What is the probability that *none* of the machines will break down during a particular shift?

The probability is $\frac{156}{275}$. The probability that a particular machine will not break down is $1 - p(\text{breakdown})$ which gives $\frac{13}{15}$, $\frac{9}{10}$, and $\frac{8}{11}$ as the three probabilities. Multiply them together and we find the probability that all three machines will fail to break down (that is, that none of them will break down):

$$\frac{13}{15} \times \frac{9}{10} \times \frac{8}{11} = \frac{156}{275}$$

You've now had quite a bit of practice with our two rules for combining probabilities—the addition rule and the multiplication rule. See if you can apply either or both of them to the following question.

Eggs for sale at a grocer's shop are graded into three sizes. Only two farms supply the grocer with eggs, and for a particular week their supplies were graded as follows:

	Total number of eggs	Percentage		
		Large	Medium	Small
Farm A	2000	20	30	50
Farm B	3000	40	30	30

If a customer chooses an egg at random from the week's total supply, calculate

the probability that it would be graded
(i)   large,
(ii) either medium or small.
The grocer actually combines the supplies within each grading when offering
the eggs for sale. If a customer then chooses one egg at random from each of the
three grades, find the probability that all three eggs were supplied by farm B.

You probably started on this problem by converting the percentages into
actual numbers of eggs:

	Total number of eggs	Number of eggs		
		Large	Medium	Small
Farm A	2000	400	600	1000
Farm B	3000	1200	900	900
Totals	5000	1600	1500	1900

(i) 1600 of the 5000 eggs are large. So the probability that a randomly chosen
egg would be large is $\frac{1600}{5000} = 0.32$.

(ii) If the randomly-chosen egg is not large then it must be either medium
or small. So, since $p + q = 1$, the probability of choosing an egg that is
medium or small must be $1 - 0.32 = 0.68$. (Alternatively, using the addition
law, we can combine the probability of choosing a medium egg $(\frac{1500}{5000})$ with
the probability of choosing a small egg $(\frac{1900}{5000})$ to get the total probability
of choosing one or the other; and $\frac{1500}{5000} + \frac{1900}{5000} = \frac{3400}{5000} = 0.68$, as before.)

Finally, we are looking for the joint probability of three separate results all
happening—the customer choosing a farm **B** egg from each of the three
grades. The probability that the large egg he chooses will be from farm **B** is
$\frac{1200}{1600}$; similarly, $p(\text{medium, farm B}) = \frac{900}{1500}$, and $p(\text{small, farm B}) = \frac{900}{1900}$.

We combine these separate probabilities by the multiplication rule:
$$\frac{1200}{1600} \times \frac{900}{1500} \times \frac{900}{1900} = \frac{3}{4} \times \frac{3}{5} \times \frac{9}{19} = 0.21$$

When Results Are NOT Independent

You should now feel confident about using the multiplication rule to calculate
the probability of several separate results *all* happening. You simply multiply
together their separate probabilities. Now for a slight complication. In every

case we've looked at so far, the results have been not only separate but also *independent.* That is to say, the probability of any one of the results did *not* depend on whether one of the other results happened.

But you'll often need to calculate the joint probability of separate results that are *not* independent. In such cases, the probability of any given result will differ according to which other results are taking place. For instance, suppose I have a bag containing 2 black marbles and 1 white marble. I put in my hand, draw out a marble at random, and throw it away. Then I put in my hand again and draw out another marble.

What is the probability that this second marble will be *black*?

Is it (a) $\frac{1}{2}$?

or (b) 1?

or (c) it all depends?

The correct answer is *(c) it all depends.* The probability of choosing a black marble the second time depends on what I chose first time. If I chose a black marble first time, then one of the two remaining marbles will be black. So the probability of choosing a black marble second time will be $\frac{1}{2}$. But if I chose the white marble first time, then both remaining marbles will be black. So I would be certain to choose a black one second time: $p = 1$. In short, the probability of this result depends on another result.

Fortunately, we use the same multiplication rule to combine such probabilities. We still multiply the separate probabilities together to find the total probability of them all happening. But we have to take care to reckon the separate probabilities of the second (and later) results *as if* the earlier ones have already happened.*

Suppose we want to calculate the probability of choosing first a white marble and then a black marble from the bag containing (to *begin* with) 2 black and 1 white. The probability of a white at the first attempt is $\frac{1}{3}$. Once a white has been chosen (but not replaced), the probability of a black, second time, will be $\frac{2}{2} = 1$. So the joint probability of choosing a white marble followed by a black marble will be: $\frac{1}{3} \times 1 = \frac{1}{3}$.

What is the probability of choosing first a black marble and then a white marble? (Remember that the first marble is thrown away after it is chosen.)

---

*Even if all the results are happening at once, it's helpful to think of them happening a set order. But, as you'll soon see, it doesn't matter which order—the joint probability will work out the same.

The probability of choosing a black marble first is $\frac{2}{3}$. The probability of then choosing a white marble is $\frac{1}{2}$. So the joint probability of choosing first a black marble and then a white is $\frac{2}{3} \times \frac{1}{2} = \frac{1}{3}$. (As you see, the joint probability is the same whichever marble is chosen first.)

Try this one:
There are five people in a doctor's waiting-room: 2 men and 3 women. What is the probability that the first two people called in to see the doctor will be a man and a woman in that order?

p(man then woman) $= \frac{3}{10}$ There are 2 men and 3 women. So the probability of a man being called in first is $\frac{2}{5}$. But, once he has been called in, there are only 4 people left in the waiting-room, 3 of them women. So the probability of a woman being called in next is $\frac{3}{4}$ (NOT $\frac{3}{5}$). So the combined (joint) probability is $\frac{2}{5} \times \frac{3}{4} = \frac{3}{10}$.

Try another example:
In a pack of playing cards there are 13 hearts. So the chance of drawing a heart is $\frac{13}{52}$ or $\frac{1}{4}$. All right, suppose you draw one card from the pack and keep it in your hand. Then you draw a second card.
    What is the probability that *both* your cards are hearts?

The probability that you have drawn two hearts is $\frac{1}{17}$. The chance of drawing a heart on the second choice was not the same as for the first choice. With 52 cards the chance of drawing a heart was $\frac{13}{52}$ or $\frac{1}{4}$. But, if the first card chosen *was* a heart, then this left only 12 hearts and 52 cards altogether in the pack. (The first card was not replaced and so was not available for the second choice.) So the probability of drawing a heart second time (if you've already drawn one on the first choice) is $\frac{12}{51}$. We then use the multiplication rule to calculate the joint probability of a heart on the first draw followed by a heart on the second: $\frac{13}{52} \times \frac{12}{51} = \frac{1}{4} \times \frac{4}{17} = \frac{1}{17}$.

What is the probability of drawing *three* hearts in a row if the cards are *not* replaced in the pack once they've been chosen?

Probability (three hearts) $= \dfrac{13}{52} \times \dfrac{12}{51} \times \dfrac{11}{50} = \dfrac{11}{850}$

With or Without Replacement? Clearly, this is a vital question in dealing with the probabilities of choosing. If the chosen items are not replaced, the probabilities change with each choice. Here is another example.

A box contains 6 red balls and 4 white ones. What is the probability of drawing out (at random) three white balls in succession, if:
  (i)   each ball is replaced before the next draw?
  (ii) the balls are not replaced?

(i) p(white) $= \dfrac{4}{10} = \dfrac{2}{5}$ each time. Probability of all three balls chosen being

white $= \dfrac{2}{5} \times \dfrac{2}{5} \times \dfrac{2}{5} = \dfrac{8}{125}$

(ii) p(white) $= \dfrac{4}{10}$ for first draw

$= \dfrac{3}{9}$ for second draw

$= \dfrac{2}{8}$ for third draw

p(three whites) $= \dfrac{4}{10} \times \dfrac{3}{9} \times \dfrac{2}{8} = \dfrac{24}{720} = \dfrac{1}{30}$

As you can see, then, in choosing two or more items from a small group, it makes quite a difference to the probabilities if the items are not replaced between choices. They will *not* be independent. Always be on the lookout for this when using the multiplication rule.

Let's have one last dip in that box containing the 6 red and 4 white balls. What is the probability of drawing out the balls in the order red-white-red-white etc., until only red balls (two of them) are left in the box? (Draw a diagram if it will help.)

Your calculation should go like this:

$$\text{(R)} \quad \text{(W)} \quad \text{(R)} \quad \text{(W)} \quad \text{(R)} \quad \text{(W)} \quad \text{(R)} \quad \text{(W)}$$

$$\dfrac{6}{10} \times \dfrac{4}{9} \times \dfrac{5}{8} \times \dfrac{3}{7} \times \dfrac{4}{6} \times \dfrac{2}{5} \times \dfrac{3}{4} \times \dfrac{1}{3} = \dfrac{1}{120}$$

The total number of balls, and the number of red or white balls, reduces by one with each choice, so the fraction changes each time.

And now a game with cards that could keep you occupied for some while. Each letter of the alphabet is written on a separate card, and the 26 cards are thoroughly shuffled. What is the probability of drawing cards marked C, A, R, D, S, in that order, if you draw five cards without replacing between draws? (Don't work the answer right out—just show the separate probabilities and how they have to be combined.)

$$p(\text{C, A, R, D, S}) = \frac{1}{26} \times \frac{1}{25} \times \frac{1}{24} \times \frac{1}{23} \times \frac{1}{22}$$

(In case you're curious, this comes to $\frac{1}{7\,893\,600}$. And one way of looking at this figure is to say that if you could shuffle and deal out a new set of five cards every 15 seconds and you worked at it 8 hours a day, every day of the week, you might come up with the right permutation about once every nine years!)

So, whenever, a problem involves a series of choices from a group of items, watch out to see whether the items are replaced between choices. If the group is a small one (say less than 50 items), the probabilities will change significantly from choice to choice if the items are not replaced. But if the group is a *large* one, it won't make much practical difference whether they are replaced or not. Thus, for instance: If you had a box containing 600 red balls and 400 white, your chances of drawing three white balls in succession would be about __?__, regardless of whether you replaced them between choices or not.

... about $\frac{8}{125}$. If you compare the products $\frac{400}{1000} \times \frac{400}{1000} \times \frac{400}{1000}$ and $\frac{400}{1000} \times \frac{399}{999} \times \frac{398}{998}$, you'll see there's little difference.

Try these examples:
(i) A machine produces components 15% of which are defective. If 3 components are selected at random, what is the probability that they will include *no defectives*?
(ii) A tray of 20 components contains 3 defectives. If 3 components are selected at random, what is the probability that they will include no defectives?

(i) Here we can assume we are dealing with such a large number of components that the probability of choosing a defective is not influenced by earlier choices. The probability of choosing a defective is $\frac{15}{100}$. But we need three non-defectives. The probability of choosing a non-defective is $1 - \frac{15}{100} = \frac{85}{100}$. The joint probability of choosing three such components is:

$$\frac{85}{100} \times \frac{85}{100} \times \frac{85}{100} = \frac{614\,125}{1\,000\,000} = 0.614$$

(ii) Here we are dealing with small numbers. So, if an item is chosen and not replaced, the probability of choosing later items is altered. The probability of choosing one of the 17 non-defectives at the first attempt is $\frac{17}{20}$. If that component is not replaced, there will be 16 non-defectives left among the 19 remaining components. So the probability of choosing a non-defective at the second attempt is $\frac{16}{19}$. Similarly, the probability of the third component being non-defective is $\frac{15}{18}$. The joint probability of all three components being non-defective is:

$$\frac{17}{20} \times \frac{16}{19} \times \frac{15}{18} = \frac{4080}{6840} = 0.596$$

So often with probability problems, the arithmetic is simple enough—once you have decided just what the question is asking for. In other words, you need to think very *clearly* about the situation you are dealing with. Never jump to conclusions; you may be overlooking something (e.g. whether or not items are replaced between choices).

Have a look at this problem:

In a single throw of two dice, what is the probability that the total will be an even number and that a 6 will appear on just one of the dice?

Will the multiplication rule give us the answer? If yes, how? If not, why not?

The multiplication rule will *not* give us the answer here. You'll remember that the rule requires that the two (or more) results be *separate*. These two results (an even total and a 6 on just one die) are NOT separate. Some of the outcomes that would give an even total would also have given a 6 on one die, e.g. $6 + 2$, $4 + 6$, etc. So one of the results is included in the other.

If we had tried to use the multiplication rule, what would have happened? Looking back at the table of possible outcomes of throwing two dice (page 35) you would have seen that eighteen of the thirty-six possible totals are even. Thus $p(\text{even}) = \frac{18}{36} = \frac{1}{2}$. Similarly, you would have noticed that ten of the thirty-six outcomes showed a 6 on one die only. Thus $p(\text{one } 6) = \frac{10}{36}$. But it would have been quite WRONG to multiply these probabilities and get $\frac{1}{2} \times \frac{10}{36} = \frac{5}{36}$. In fact, it is clear that only four outcomes would give the results we are looking for: 6 on the first die together with 2 or 4 on the second; or 6 on the second with 2 or 4 on the first. So, only four out of thirty-six equally likely outcomes give the results we are looking for—an even total and a 6 on just one die (i.e. $6 + 2$, $6 + 4$, $2 + 6$ and $4 + 6$). So the probability is $\frac{4}{36} = \frac{1}{9}$.

The multiplication rule would have given the wrong answer here because the results were not really separate. One was contained in the other. Never apply the rule unless you are sure the results you are interested in are quite separate (but not mutually exclusive). They may or may not be independent also. (As a matter of fact, there is a way of tackling that last problem by using the multiplication and addition rules *together*. You'll know how to do this after you've been through the next section.)

By the way, can two results be both mutually exclusive *and* independent? Yes or No? Why or Why Not?

No, two results cannot be both mutually exclusive and independent. If they are mutually exclusive, each prevents the other from happening. Therefore, each depends on the other. They are not independent.

Now, to finish this section, try this question. Four boys are taken at random. What is the probability that they were:
(i) all born on the same day of the week,
(ii) all born on different days of the week?
It may be assumed that births are equally likely on the seven days of the week.

(i) Let's call the four boys A, B, C, and D. What we have to calculate is the joint probability that *three* boys (say B, C, and D) were all born on the same day as A. There is a $\frac{1}{7}$ probability that B was born on that day. Similarly for C and D. So the joint probability of all four boys being born on the same day is:

$$\frac{1}{7} \times \frac{1}{7} \times \frac{1}{7} = \frac{1}{343}$$

(ii) This time the probabilities are not independent. Whatever day A was born, B has a $\frac{6}{7}$ chance of being born on a different day. However, two of the seven days have now been 'used up', so C's chance of being born on a different day from A and B is only $\frac{5}{7}$. And D has only four days left to 'choose' from; so his probability of a different birthday is down to $\frac{4}{7}$. So the probability that the four boys will all be born on different days is:

$$\frac{6}{7} \times \frac{5}{7} \times \frac{4}{7} = \frac{120}{343}$$

Now that you can use the multiplication rule, as well as the addition rule, we'll concentrate, in the next section, on problems that require you to use both rules *together*.

Using Both Probability Rules Together

You can now combine probabilities in two different ways:
1 By adding........ (Total probability) ... addition rule
2 By multiplying ... (Joint probability) ... multiplication rule.
Many problems in probability, however, will require you to use both rules together. (And you'll need to think carefully to see what is called for.)

Suppose we are playing a game with a bucket containing eight balls—five black and three white. What is the probability of your winning if what you have to do is take out (blindfold) one black ball and then one white ball, in that order and without replacement?

p(black ball then white) $= \frac{15}{56}$. The chance of choosing a black ball first is $\frac{5}{8}$

and, when that has been removed, the chance of choosing a white ball is $\frac{3}{7}$. So

the combined probability is $\frac{5}{8} \times \frac{3}{7} = \frac{15}{56}$.

So, if you could win the game by selecting a black ball followed by a white ball, your chance of doing so would be $\frac{15}{56}$. Suppose the rules of the game are changed—instead of black followed by white, you must now choose white followed by black.

What would be your chance of picking out a white ball first and then a black, given the same eight balls as before (and no replacement)?

$p = \frac{15}{56}$, exactly as before. (That is, $\frac{3}{8} \times \frac{5}{7} = \frac{15}{56}$.)

So your chance of winning the game would be no different. You are as likely to get a black and a white as a white and a black.

Now suppose the rules were changed yet again, so that either result would be counted as a win: EITHER black followed by white OR white followed by black.

How would this change affect your chances of winning the game? They would now be ...

(a) smaller than $\frac{15}{56}$?

or (b) larger than $\frac{15}{56}$?

or (c) exactly $\frac{15}{56}$?

(b) larger than $\frac{15}{56}$. You are now being given two equally likely ways to win the game instead of one; therefore you have more chance. You can win by drawing:
EITHER a black ball then a white ball (Result 1)
    OR a white ball then a black ball (Result 2)

Would you describe these two possible results as *independent* or as *mutually exclusive*?

The two possible results are mutually exclusive. If you've drawn a black ball and then a white one you cannot, at the same time, have drawn a white then a black (and vice versa).

So here we have two mutually exclusive results with the probabilities we have calculated:

$$\text{EITHER } p(\text{Black then White}) = \frac{15}{56}$$

$$\text{OR } p(\text{White then Black}) = \frac{15}{56}$$

What is your probability of winning the game by drawing one sequence or the other?

$p(\text{B then W or W then B}) = \frac{15}{28}$. Since the two results are mutually exclusive you can use the addition rule to combine the separate probabilities. Thus, in solving the problem, we have used both the multiplication rule and the addition rule:

$p(\text{B then W})$		$p(\text{W then B})$		$p(\text{EITHER B then W OR W then B})$
$\frac{5}{8} \times \frac{3}{7} = \frac{15}{56}$	$+$	$\frac{3}{8} \times \frac{5}{7} = \frac{15}{56}$	$=$	$\frac{30}{56} = \frac{15}{28}$

↑multiplication   ↑multiplication
rule   rule
        addition
        rule

So, given the choosing of two balls from a bucket containing 5 black balls and 3 white ones, we've considered two possible joint results—black ball followed by white OR white ball followed by black.

But there are two other possible joint results. What are they?
        __?__ ball followed by __?__ ball
    and __?__ ball followed by __?__ ball.

The other two possible joint results are:
   black ball followed by *black* ball
and *white* ball followed by *white* ball.

Remember, there are 5 black balls and 3 white balls.
What is the probability of choosing (without replacement):
   (i) two black balls?
   (ii) two white balls?

The probability of choosing

(i) two black balls is $\dfrac{5}{8} \times \dfrac{4}{7} = \dfrac{20}{56}$

(ii) two white balls is $\dfrac{3}{8} \times \dfrac{2}{7} = \dfrac{6}{56}$

The two possible joint results of a double draw are mutually exclusive—if you've drawn two blacks you clearly can't have drawn two whites (and vice versa).

   So what is the probability that you will draw EITHER two black balls OR two white ones?

$p(\text{EITHER B then B OR W then W}) = \dfrac{26}{56}$. Since the two possible joint results

were mutually exclusive, we used the addition rule to combine their probabilities. That is, having used the multiplication rule to find the probabilities of the two separate results:

$$p(\text{B then B}) = \frac{5}{8} \times \frac{4}{7} = \frac{20}{56}$$

$$\text{and } p(\text{W then W}) = \frac{3}{8} \times \frac{2}{7} = \frac{6}{56}$$

we used the addition rule to find the probability that either one or the other of these results will occur:

$$p(\text{Either B then B or W then W}) = \frac{20}{56} + \frac{6}{56} = \frac{26}{56}.$$

Now, to finish with this problem, here are the four joint probabilities we calculated using the multiplication rule:

$$p(\text{B then W}) = \frac{15}{56}$$

$$p(\text{W then B}) = \frac{15}{56}$$

$$p(\text{B then B}) = \frac{20}{56}$$

$$p(\text{W then W}) = \frac{6}{56}$$

What is the probability that either one or another of these four possible results will take place?

Since the four possible results are mutually exclusive we can use the addition rule to calculate the probability that one or another of them will take place:

$$\frac{15}{56} + \frac{15}{56} + \frac{20}{56} + \frac{6}{56} = \frac{56}{56} = 1$$

which is what you would expect since there are no other possible results and one of the four (B then W, or W then B, or B then B, or W then W) is *certain* to happen.

Now let's have a change of scene. Another example where you must first use the multiplication rule and then the addition rule.

What is the probability of throwing a head and a tail (any order) in tossing two coins?

---

p(HT or TH) $= \dfrac{1}{2}$. There are two ways you can throw a head and a tail:

$$\left.\begin{array}{l} p(\text{H then T}) = \dfrac{1}{2} \times \dfrac{1}{2} = \dfrac{1}{4} \\[2mm] \text{or} \\[1mm] p(\text{T then H}) = \dfrac{1}{2} \times \dfrac{1}{2} = \dfrac{1}{4} \end{array}\right\} \; p(\text{HT or TH}) = \dfrac{1}{4} + \dfrac{1}{4} = \dfrac{1}{2}$$

Once again, we have considered two mutually exclusive results (head followed by tail OR tail followed by head) and have thus been able to use the addition rule to calculate the *total* probability. You can check this result by looking at the table of equally likely outcomes you drew up earlier (page 33) or at your probability tree.

What is the probability of getting *two* heads and one tail in *three* throws of a coin (any order)?

$p = \dfrac{3}{8}$ for two heads and a tail (in any order).

Since $p(\text{H}) = \dfrac{1}{2}$ and $p(\text{T}) = \dfrac{1}{2}$:

$$p(\text{HHT}) = \frac{1}{2} \times \frac{1}{2} \times \frac{1}{2} = \frac{1}{8}$$

$$p(\text{HTH}) = \frac{1}{2} \times \frac{1}{2} \times \frac{1}{2} = \frac{1}{8}$$

$$p(\text{THH}) = \frac{1}{2} \times \frac{1}{2} \times \frac{1}{2} = \frac{1}{8}$$

$$\text{Total} = \frac{3}{8}$$

Again you can check this from the probability tree you drew up earlier (page

37). Clearly, you would have exactly the same probability of getting two *tails* and a head in any order.

When working through this kind of problem, we have to be very systematic to make su.e we don't miss any of the possible joint results. Let's try it again with a new example.

We have a box containing a dozen eggs, three of which are bad. If we select three eggs at random to make an omelette, what is the probability they will all be bad?

$p = \dfrac{6}{1320}$ that all three eggs will be bad. Since the eggs selected are not being replaced, the chance of choosing a bad egg changes from choice to choice:

$p(\text{1st choice}) = \dfrac{3}{12}$

$p(\text{2nd choice}) = \dfrac{2}{11}$ To calculate these probabilities we have assumed that the previous choice gave a bad egg. (We are not inter-

$p(\text{3rd choice}) = \dfrac{1}{10}$ ested in choices that might follow the choosing of a good egg.)

We use the multiplication rule to find the probability of getting a run of three bad eggs:

$$p(\text{all 3 eggs bad}) = \dfrac{3}{12} \times \dfrac{2}{11} \times \dfrac{1}{10} = \dfrac{6}{1320}$$

So, with three bad eggs in a dozen, there's not much chance we'd pick all bad eggs if we chose three at random.

Now what is the probability of selecting NO bad eggs at all if we take three at random?

$p = \dfrac{504}{1320}$ (that all three eggs are good). The probability is found by multiplying: $\dfrac{9}{12} \times \dfrac{8}{11} \times \dfrac{7}{10} = \dfrac{504}{1320}$.

All right, we have twelve eggs, three of which are bad. So far we've considered two possible ways of choosing three eggs from them:

$$p(\text{all 3 eggs bad}) = \dfrac{6}{1320}$$
$$p(\text{all 3 eggs good}) = \dfrac{504}{1320}$$

So, on average, we might expect to pick either all bad or all good eggs on about 510 out of every 1320 occasions. Now what about all those other 810 occasions (1320 − 510) on which we could expect to pick some good, some bad?

What is the probability of selecting 1 bad egg, and 2 good ones? (Work out all the ways this combination could happen.)

In selecting 3 eggs from 12, 3 of which are bad:

$$p(\text{1st bad, other two good}) = \frac{3}{12} \times \frac{9}{11} \times \frac{8}{10} = \frac{216}{1320}$$

$$p(\text{2nd bad, other two good}) = \frac{9}{12} \times \frac{3}{11} \times \frac{8}{10} = \frac{216}{1320}$$

$$p(\text{3rd bad, other two good}) = \frac{9}{12} \times \frac{8}{11} \times \frac{3}{10} = \frac{216}{1320}$$

So there are three alternative ways you could get one bad egg in a selection of three, and each possible way has a probability of $\frac{216}{1320}$.

How do we combine these probabilities to find the total probability of getting one bad egg and two good ones?

    (a) add them?

or (b) multiply them?

We *add* the probabilities. The bad egg would be either first, second, or third. If it was in one position, it couldn't be in any of the others. So the three possible outcomes (BGG, or GBG, or GGB) are mutually exclusive. And the probability of getting one bad egg and two good eggs is (as you may already have calculated):

$$\frac{216 + 216 + 216}{1320} = \frac{648}{1320}$$

Here are the probabilities we have calculated so far:

$$p(\text{3 bad eggs}) \quad = \frac{6}{1320}$$

$$p(\text{3 good eggs}) \quad = \frac{504}{1320}$$

$$p(\text{1 bad, 2 good}) = \frac{648}{1320}$$

Is there any other combination of good and bad eggs that we have not yet considered? If so, what?

Yes, we've not yet considered the combination of 2 bad and 1 good egg.

So far we've had these selections:

1st choice	2nd choice	3rd choice		
Bad	Bad	Bad	→ 3 bad	$p = \dfrac{6}{1320}$
Good	Good	Good	→ 3 good	$p = \dfrac{504}{1320}$
Bad	Good	Good		
Good	Bad	Good	→ 1 bad, 2 good	$p = \dfrac{648}{1320}$
Good	Good	Bad		

Now write down the sequences of good and bad that we have *not* yet considered.

The sequences we have not yet considered are:

$$\left.\begin{array}{lll} \text{Good} & \text{Bad} & \text{Bad} \\ \text{Bad} & \text{Good} & \text{Bad} \\ \text{Bad} & \text{Bad} & \text{Good} \end{array}\right\} \rightarrow \text{2 bad, 1 good } p = ?$$

Remembering that there are 9 good eggs in the dozen, would you expect the probability of picking 2 bad and 1 good to be higher or lower than the probability of picking 1 bad and 2 good?

Now to check whether you are right: calculate the probability of selecting 2 bad eggs and 1 good one.

The probabilities for the three ways of choosing 2 bad eggs and 1 good one are:

$$p(\text{GBB}) = \frac{9}{12} \times \frac{3}{11} \times \frac{2}{10} = \frac{54}{1320}$$

$$p(\text{BGB}) = \frac{3}{12} \times \frac{9}{11} \times \frac{2}{10} = \frac{54}{1320}$$

$$p(\text{BBG}) = \frac{3}{12} \times \frac{2}{11} \times \frac{9}{10} = \frac{54}{1320}$$

So the total probability of selecting either GBB or BGB or BBG is:

$$\frac{54 + 54 + 54}{1320} = \frac{162}{1320}$$

(Which, as you no doubt expected, is *less* than the probability of selecting 1 bad egg and 2 good.)

Here, then, are all the probabilities we have calculated:

$$p(\text{3 bad eggs}) \quad = \frac{6}{1320}$$

$$p(\text{3 good eggs}) \quad = \frac{504}{1320}$$

$$p(\text{1 bad, 2 good}) = \frac{648}{1320}$$

$$p(\text{2 bad, 1 good}) = \frac{162}{1320}$$

What is the SUM of these separate probabilities?

As you would expect, the sum is 1:

$$\frac{6 + 504 + 648 + 162}{1320} = \frac{1320}{1320} = 1$$

That is, in selecting three eggs from twelve (three of which are bad) we MUST get *either* 3 bad *or* 3 good *or* 1 bad and 2 good *or* 2 bad and 1 good. These alternative results are mutually exclusive and there are no other possible combinations; so their separate probabilities must add up to 1—absolute certainty. (And, of course, this fact gives us a useful check on all the arithmetic we've been doing.)

Now try this question.
A school bicycle shed contains 25 bicycles, seven of which have defective brakes. If three bicycles are selected at random for testing, what is the probability that exactly one will have defective brakes?

The probability that the first bicycle has defective brakes and the second and third have sound brakes is:

$$\frac{7}{25} \times \frac{18}{24} \times \frac{17}{23} = \frac{2142}{13\,800}$$

Similarly, the probability that the second bicycle has defective brakes while the other two have sound brakes is:

$$\frac{18}{25} \times \frac{7}{24} \times \frac{17}{23} = \frac{2142}{13\,800}$$

And the probability that the third bicycle is the only one with defective brakes is:

$$\frac{18}{25} \times \frac{17}{24} \times \frac{7}{23} = \frac{2142}{13\,800}$$

So there are three ways we could get exactly one bicycle with defective brakes—it must be either the first or the second or the third. The total probability is therefore:

$$\frac{2142 + 2142 + 2142}{13\,800} = \frac{6426}{13\,800} = 0.466$$

At this point you should be able to take a new approach to a problem we last looked at on page 49.

In a single throw of two dice, what is the probability that the total will be an even number and that a 6 will appear on just one of the dice?

We decided, you'll remember, that we couldn't solve this problem using the multiplication rule alone. Try to solve it now using the multiplication and addition rules *together*.

The solitary 6 must appear on either the first die or the second. (It can't be on both.) And, for the total to be an even number, the *other* die must score 2 or 4. (But not 6, because 6 is to appear on one die only.)

The probability of 6 on the *first* die is $\frac{1}{6}$. The probability of 2 or 4 (two of six equally likely outcomes) on the second die is $\frac{2}{6}$. So, by the multiplication rule, the joint probability of this sequence is $\frac{1}{6} \times \frac{2}{6} = \frac{2}{36}$. Similarly, the probability

of 2 or 4 on the first die, with 6 appearing on the *second* die, is $\frac{2}{6} \times \frac{1}{6} = \frac{2}{36}$.
These are mutually exclusive results—either is possible, but not both. So the total probability of one sequence or the other happening is found by the addition rule: $\frac{2}{36} + \frac{2}{36} = \frac{4}{36} = \frac{1}{9}$. (This, as you'll recall, is the same answer we decided on last time we tackled the problem.)

One Only v. At Least One

There is an important difference in probability, as in everyday conversation, between the phrases 'one only' and 'at least one'. That last problem you tackled was a case of 'one only'—just one die showing a 6. Similarly with the previous problem—exactly one of the bicycles to have defective tires. Let's try some more of these 'one only' problems before we go on to see how they differ from 'at least one' probabilities.

Suppose you are to draw two cards, one from each of two packs. What is the probability that one, and *one only*, will be a diamond? With two choices, there are *two* ways you could get the diamond you need. You could draw:
EITHER: Diamond from 1st pack, something else from 2nd pack:

$$p = \frac{1}{4} \times \frac{3}{4} = \frac{3}{16}$$

OR: Something else from 1st pack, diamond from 2nd pack:

$$p = \frac{3}{4} \times \frac{1}{4} = \frac{3}{16}$$

So what is the total probability that you will draw one diamond only?

---

p(one diamond only) $= \dfrac{6}{16}$

The total probability of choosing one diamond only is the sum of the *two* separate ways of doing it:

$$\begin{array}{ccc} \text{diamond} & & \text{diamond} \\ \text{1st time} & \text{OR} & \text{2nd time} \\ \left(\frac{1}{4} \times \frac{3}{4}\right) & + & \left(\frac{3}{4} \times \frac{1}{4}\right) = \frac{6}{16} \end{array}$$

These two ways of getting one diamond are mutually exclusive.

Now try an example where there are *three* ways of getting the *one* result we need:

A local garage has three breakdown trucks capable of hauling away damaged vehicles. For various reasons, the probability that a particular truck will be available when needed is $\frac{9}{10}$. The availability of each truck is independent of that of the others.

If you telephone the garage now, what is the probability that one, and only one, of the trucks will be available? (Call them A, B and C.)

If one truck only is to be available, it must be A or B or C. So there are three ways of getting one truck available:

$$p(\text{A available, others not}) = \frac{9}{10} \times \frac{1}{10} \times \frac{1}{10} = \frac{9}{1000}$$

$$p(\text{B available, others not}) = \frac{1}{10} \times \frac{9}{10} \times \frac{1}{10} = \frac{9}{1000}$$

$$p(\text{C available, others not}) = \frac{1}{10} \times \frac{1}{10} \times \frac{9}{10} = \frac{9}{1000}$$

And, since these three alternative ways are mutually exclusive, the probability of one truck only being available is found by the addition rule:

$$p(\text{A or B or C}) = \frac{9 + 9 + 9}{1000} = \frac{27}{1000}$$

Another example.
Three archers (Rob, Will and John) are taking part in a shooting contest. Judging by recent performance, their chances of hitting the bull's-eye with one arrow are estimated as follows:

Rob: $\frac{1}{2}$ Will: $\frac{1}{3}$ John: $\frac{1}{4}$

Estimate the chance that *one, and only one*, archer will hit the bull's-eye on his target.

What we are looking for is the probability that one of the men will hit while the other two miss. This can happen in one of three ways:

$$p(\text{1st man hits, other two miss}) = \frac{1}{2} \times \frac{2}{3} \times \frac{3}{4} = \frac{6}{24}$$

OR

$$p(\text{2nd man hits, other two miss}) = \frac{1}{2} \times \frac{1}{3} \times \frac{3}{4} = \frac{3}{24}$$

OR

$$p(\text{3rd man hits, other two miss}) = \frac{1}{2} \times \frac{2}{3} \times \frac{1}{4} = \frac{2}{24}$$

These three possible results are mutually exclusive, so the sum of their separate probabilities gives us the probability that one or other will happen:

$$p(\text{one only}) = \frac{6 + 3 + 2}{24} = \frac{11}{24}$$

Now work out the probability that:
(i) two of the archers* hit the bull's-eye

*For this to happen, one of the archers must miss; so again there are three ways it can happen.

(ii) all three of them hit

(iii) none of them hit.

(i) p(1st man misses, other two hit) $= \dfrac{1}{2} \times \dfrac{1}{3} \times \dfrac{1}{4} = \dfrac{1}{24}$

   p(2nd man misses, other two hit) $= \dfrac{1}{2} \times \dfrac{2}{3} \times \dfrac{1}{4} = \dfrac{2}{24}$

   p(3rd man misses, other two hit) $= \dfrac{1}{2} \times \dfrac{1}{3} \times \dfrac{3}{4} = \dfrac{3}{24}$

   Combined probability (1 miss, 2 hit) $= \dfrac{1 + 2 + 3}{24} = \dfrac{6}{24}$

(ii) p(all 3 hit) $= \dfrac{1}{2} \times \dfrac{1}{3} \times \dfrac{1}{4} = \dfrac{1}{24}$

(iii) p(none hit) $= \dfrac{1}{2} \times \dfrac{2}{3} \times \dfrac{3}{4} = \dfrac{6}{24}$

And we have already worked out that

(iv) p(2 miss, 1 hit) $= \dfrac{11}{24}$

Notice that $\dfrac{6}{24} + \dfrac{1}{24} + \dfrac{6}{24} + \dfrac{11}{24} = 1$, showing that we have included all possible results and that one or other of them is *certain* to happen.

Now what is the difference between 'one only' and 'at least one'? Using the results above, what is the probability that *at least one* archer will hit the bull's-eye?

Since there are 3 archers, 'at least one' means 1, 2, or 3. So:

$$p(1 \text{ archer hits bull's-eye}) = \frac{11}{24}$$

$$p(2 \text{ archers hit bull's-eye}) = \frac{6}{24}$$

$$p(3 \text{ archers hit bull's-eye}) = \frac{1}{24}$$

These are three mutually exclusive results—only one of them can take place. So the combined probability of one or another is:

$$\frac{11 + 6 + 1}{24} = \frac{18}{24} = p(\text{at least 1 archer hits bull's-eye})$$

We have been able to use the probabilities we calculated earlier in order to find a new probability—the probability that at least one archer will hit the bull's-eye. But there is an easier way to find such a probability. After all, the figure of $\dfrac{18}{24}$ represents the combined probability of either 1, 2, or 3 archers hitting the bull's-eye. But what about the remaining

$$1 - \frac{18}{24} = \frac{6}{24}$$

This represents the probability that (*how many?*) of the archers hits the bull's-eye?

This represents the probability that *none of the archers hits the bull's-eye*. Either none of the archers hits the bull's-eye *or* at least one of them (i.e. 1, 2, or 3) does. The probability of one or the other outcome is 1. So:

$$1 - p(\text{at least one hits bull's-eye}) = p(\text{none hits bull's-eye})$$

Looking at this the other way round:

To find the probability that at least one archer hits the bull's-eye, we'd need to subtract from 1 the probability that (*your own words*).

...we'd need to subtract from 1 the probability that *none of the archers hits the bull's-eye*. As you'll recall, the archers' individual probabilities of hitting were $\frac{1}{2}, \frac{1}{3}$, and $\frac{1}{4}$ The probability that none of them would hit is therefore:

$$\frac{1}{2} \times \frac{2}{3} \times \frac{3}{4} = \frac{6}{24}$$

So, to check, the probability that at least one would hit is:

$$1 - \frac{6}{24} = \frac{18}{24}$$

which is, of course, exactly what we found by adding up the probabilities of 1, 2, and 3 hitting.

Let's try another example of 'at least one ...' probability.

A man forecasts the results of two football games. What is the probability that he EITHER gets both wrong OR gets at least one of them right?

He is certain to get one or two of them right, or get them both wrong. So the probability is 1.

If we say he has a $\frac{2}{3}$ chance of being wrong on each game, what is the probability that he gets both wrong?

His probability of getting both wrong is: $\frac{2}{3} \times \frac{2}{3} = \frac{4}{9}$

So, what is the probability he will be right at least once?

The probability he will be right at least once is:

$$1 - p(\text{wrong both times}) = 1 - \frac{4}{9} = \frac{5}{9}$$

Notice that this is greater than his chance of getting one and one only forecast right, $(\frac{1}{3} \times \frac{2}{3}) + (\frac{2}{3} \times \frac{1}{3}) = \frac{4}{9}$, because it includes the case where he gets *both* right.

At least one ... Here is how you can sum up this kind of calculation.

> The probability that AT LEAST ONE of several independent results will happen is equal to ONE MINUS the probability that NONE of them will happen.

Suppose you toss a coin *three* times. What is the probability of throwing *at least one* head?

The probability of failing to get a head at all on three throws is $\frac{1}{2} \times \frac{1}{2} \times \frac{1}{2}$ $= \frac{1}{8}$. If this is the probability of getting no heads, the probability of getting at least one is $1 - \frac{1}{8} = \frac{7}{8}$. (You could check this figure by adding up the probabilities of getting one, two, and three heads.)

You'll remember your local garage with its three breakdown trucks, each of which has a $\frac{9}{10}$ chance of being available when required. You calculated the probability of one, and one only, being available as $\frac{27}{1000}$. But what is the probability of *at least* one truck being available?

Either none of the trucks is available or at least one of them is.

$$p(\text{none available}) = \frac{1}{10} \times \frac{1}{10} \times \frac{1}{10} = \frac{1}{1000}$$
$$p(\text{at least one available}) = 1 - p(\text{none available})$$
$$= 1 - \frac{1}{1000}$$
$$= \frac{999}{1000}$$

Notice again that 'at least one' gives a bigger probability than 'one only'. This is because it contains the probability of more than one as well.

Another problem. The probabilities that three independent components in a television set will burn out within three years are, respectively, $\frac{1}{12}, \frac{1}{10}$, and $\frac{1}{11}$. Calculate the probability that *at least one* of those components will need replacing within the period.

$$p(\text{no component burns out}) = \frac{11}{12} \times \frac{9}{10} \times \frac{10}{11} = \frac{3}{4}$$

$$p(\text{at least one burns out}) \quad = 1 - \frac{3}{4} = \frac{1}{4}$$

One more:

Janet Ellis is being interviewed for three jobs. At the first interview there are three candidates altogether, at the second there are five, and at the third there are four candidates. If all candidates are equally likely to be selected, what is the probability that Janet would be offered:

(i)   one and only one of the jobs?

(ii) at least one of the jobs?

(i) If she is offered one job and one job only, it must be either the first:

$$p(\text{offered 1st job only}) \quad = \frac{1}{3} \times \frac{4}{5} \times \frac{3}{4} = \frac{12}{60}$$

or the second:

$$p(\text{offered 2nd job only}) = \frac{2}{3} \times \frac{1}{5} \times \frac{3}{4} = \frac{6}{60}$$

or the third:

$$p(\text{offered 3rd job only}) \quad = \frac{2}{3} \times \frac{4}{5} \times \frac{1}{4} = \frac{8}{60}$$

Since these are mutually exclusive events, the probability that one or the other will happen is the *sum* of the separate probabilities:

$$p(\text{offered one job only}) = \frac{12}{60} + \frac{6}{60} + \frac{8}{60} = \frac{26}{60}$$

(ii) The probability that Janet will be offered *none* of the jobs is $\frac{2}{3} \times \frac{4}{5} \times \frac{3}{4} = \frac{24}{60}$. And since she must either be offered no job or at least one of them, i.e. $p(\text{none}) + p(\text{at least one}) = 1$, then the probability of her being offered at least one job is:

$$1 - \frac{24}{60} = \frac{36}{60}$$

So, as you might have expected, she is (*more/less?*) likely to be offered at least one job than one job only.

She is *more* likely to be offered at least one job than to be offered one job only. (This is because 'at least one' carries the probabilities for more than one as well.)

If you were at all surprised by the high probability (a 3 in 5 chance) of at least one job being offered, you may well be amazed by our next example.

   Suppose you are lucky enough to have 30 friends who are likely to invite you to parties on their birthdays. You can go to one party only on any one day. What would you guess is the probability that you'll have to miss at least one party because it clashes with another birthday party on the same day of the year? (Bear in mind that the 30 birth-dates could be spread at random over the 365 days of the year.) Would you guess the probability is:

(a) $p = $ less than $\frac{1}{12}$?

or (b) $p = $ between $\frac{1}{12}$ and $\frac{1}{5}$?

or (c) $p = $ more than $\frac{1}{5}$?

Don't forget what your guess was. Let's calculate the probability now and you'll be able to see whether you were right.

What is the probability that at least one of the 30 birthdays will clash with another? Let's begin by working out the probability that *none* of the birthdays clash. Take your friends one at a time. Friend 1 has 365 days to 'choose from' for his or her birthday. This leaves 364 days for Friend 2: he or she has 364 chances out of 365 of avoiding a clash with Friend 1: i.e. $\frac{364}{365}$. Friend 3 has 363 chances out of 365 of avoiding the dates 'chosen' by Friends 1 and 2: i.e. $\frac{363}{365}$.

What is the chance that Friend 4 will avoid a clash with the previous three?

The probability that Friend 4 will avoid a clash with the previous three is 362 chances in 365 or $\frac{362}{365}$. (That is, 3 of the year's 365 days have already been taken, leaving only 362 that avoid a clash.)

Clearly, the probability of avoiding a clash becomes smaller and smaller (reducing by $\frac{1}{365}$ at each step), until we come to the 30th friend whose chance of avoiding a clash is $\frac{336}{365}$.

So each friend's probability of avoiding a clash of dates with someone earlier in the list is as follows:

1st friend	2nd	3rd	4th	probabilities for 5th to 28th friend	29th	30th friend
$\frac{365}{365}$	$\frac{364}{365}$	$\frac{363}{365}$	$\frac{362}{365}$ ⋯		⋯ $\frac{337}{365}$	$\frac{336}{365}$

What would we do to these 30 probabilities to get the probability that NONE of the dates clash?

We would multiply the probabilities together. The combined probability of a failure to clash is the product of the 30 individual probabilities:

$$\frac{365}{365} \times \frac{364}{365} \times \frac{363}{365} \times \frac{362}{365} \times \ldots \times \frac{337}{365} \times \frac{336}{365}$$

Using logarithms or a calculator, we find the product is approximately 0.3; this is the probability that none of the birthdays will fall on the same date.

So the probability of no clash is $\frac{3}{10}$. The probability of either no clash or at least one clash is, of course, 1.

So what is the probability that you will have to miss at least one of those thirty birthday parties because it falls on the same day as one of the others?

Since there must be either no clash ($p = \frac{3}{10}$) or at least one, the probability of at least one clash is $1 - \frac{3}{10} = \frac{7}{10}$.

Surprised? We have performed a perfectly genuine calculation—no tricks! Don't worry if it seems incredible. Just take it as a good example of how common sense is sometimes not quite enough when you are dealing with complex probabilities.

You might like to check this 'theoretical' probability by taking some practical samples: perhaps you can look at a class-register showing the ages of about 30 schoolchildren, or take a random selection of 30 from some publication like *Who's Who*, in which people's birth-dates are given.

If you considered a number of people smaller than 30, would you expect a clash of birth-dates to be more likely or less likely?

The table below shows how the probability of a clash becomes less likely as the number of people gets smaller:

No. of people in group	40	30	25	20	15	10	5
Approximate probability of at least one clash of birth-dates	$\frac{9}{10}$	$\frac{7}{10}$	$\frac{3}{5}$	$\frac{2}{5}$	$\frac{1}{4}$	$\frac{1}{8}$	$\frac{1}{30}$

Let's sum up the two main points of this last section. Firstly, the probability of *at least one* of several events happening is the *total* probability of one of them happening plus two of them, three of them ... through to all of them happening. Secondly, to calculate the probability that at least one of a number of independent events takes place, you subtract from 1 the probability that ..... (*complete it in your own words*).

To calculate the probability that at least one of a number of independent events takes place, you subtract from 1 the probability that *none of them takes place*.

Working 'backwards' in this way it is quite easy to calculate 'at least one' probabilities. The only trouble is that those give-away words do not always appear in the problem you are presented with. Sometimes you have to be very sharp to realise that it is indeed the total probability of one and two and three, and so on, that you should be looking for. Watch out for this from now on.

See what you think of this problem.

Two secret agents, one Russian and one Chinese, are working independently at deciphering a coded message captured from U.S. agents. The probability that the Russian will decipher it (based on his previous record) is $\frac{3}{4}$. The probability that the Chinese will decipher it is $\frac{1}{3}$. What is the probability that the coded message will be deciphered?

The code may be broken either by the Russian or by the Chinese or by both. In other words, the probability that the code will be broken is an 'at least one' probability. What is the probability that one or the other or both will decipher the message? If we'd been looking for the probability that both agents crack the code, we'd have found $p = \frac{3}{12}$. But this would not have taken account of the fact that one of the agents might crack the code while the other did not. Similarly, if we'd been interested merely in the probability that only one of the two agents would crack the code, we'd have found $p = \frac{7}{12}$, but this would be ignoring the possibility of both agents succeeding.

So, the probability that the message will be deciphered is equal to $1 - p$(message not deciphered). Since the Russian has a probability of $\frac{3}{4}$ and the Chinese a $\frac{1}{3}$ probability of deciphering it, the probability that neither will succeed is $\frac{1}{4} \times \frac{2}{3} = \frac{2}{12}$. So the probability that at least one of them (one or other or both) will succeed is $1 - \frac{2}{12} = \frac{10}{12} = \frac{5}{6}$.

If you didn't realise, at first, that this was an 'at least one' probability, take warning and be on your guard in future. So often with statistics problems, the pitfalls lie, not in the arithmetic, but in how clearly you can think out the situation suggested by the problem.

Let's try another problem. An oil company is drilling a number of holes in a certain area in the hope of finding a productive well. The probability of success on any given trial is $\frac{1}{5}$. What is the probability that:

(i)   the drilling crew are 'third time lucky' and the third hole is the first to yield oil?

(ii)   none of the first three holes yields oil?

(iii) at least one of the three holes yields oil?

(i) The probability of 'third time lucky' is: $\dfrac{4}{5} \times \dfrac{4}{5} \times \dfrac{1}{5} = \dfrac{16}{125}$

(ii) The probability of no oil in three tries is: $\dfrac{4}{5} \times \dfrac{4}{5} \times \dfrac{4}{5} = \dfrac{64}{125}$

(iii) Either you strike oil on none of the three trials or you strike it at least once. The probability of not striking oil is $\dfrac{64}{125}$, so the probability of striking it at least once is: $1 - \dfrac{64}{125} = \dfrac{61}{125}$.

Here's another. Insurance companies base their premiums for life insurance on what are called 'actuarial' calculations. Essentially, these predict people's likely life-expectancy, taking into account such factors as their age and sex. The probability that a certain man will still be alive in 25 years is $\dfrac{4}{7}$. The probability that his wife will still be alive is $\dfrac{3}{4}$. What is the probability that, in 25 years' time ...

(i)   both will be alive?

(ii)   only the man will be alive?

(iii) only the wife will be alive?

(iv)   they will not both be dead?

(i) $p(\text{both alive})$ $= \dfrac{4}{7} \times \dfrac{3}{4} = \dfrac{12}{28}$

(ii) $p(\text{man alive only}) = \dfrac{4}{7} \times \dfrac{1}{4} = \dfrac{4}{28}$

(iii) $p(\text{wife alive only}) = \dfrac{3}{7} \times \dfrac{3}{4} = \dfrac{9}{28}$

(iv) The probability that they will not both be dead is the same as the probability that either one or the other or both of them will still be alive—in short, the probability that *at least one* of them is alive. However, we don't need come at it 'backwards' by the $1 - p = q$ method this time. After all, we have already calculated the probability that the man will be alive, the probability that the woman will be alive, and the probability that both will be alive. So we simply add together the results shown in (i), (ii), and (iii) above:

$$p(\text{at least one alive}) = \dfrac{12}{28} + \dfrac{4}{28} + \dfrac{9}{28} = \dfrac{25}{28}$$

(Check by the other method if you like.)

Now calculate the probability that at least one of the couple will be dead.

Either both the man and wife are alive or at least one of them is dead. That is, p(both alive) + p(at least one dead) = 1.
So, p(at least one dead) = 1 − p(both alive)
$$= 1 - \frac{3}{7} = \frac{4}{7}$$
Remember, in this kind of problem you are using your knowledge that all the mutually exclusive results added together must give a total probability of 1.

If you shop at our local supermarket on Saturday morning, your probability of having to wait five minutes or more at a check-out is $\frac{1}{9}$. Next week my wife and I are going to try shopping separately in the supermarket and each going to check out with a different cashier. If we both reach our check-outs at the same time, what is the probability that . . .
  (i)   I will be through the check-out in less than five minutes?
  (ii) we will both be checked out in less than five minutes?
  (iii) one or other of us, or both, will be kept waiting for five minutes or more?

(i) p(me < 5 min) = $\frac{8}{9}$
The probability of only my getting through in less than 5 minutes would have been $\frac{8}{9} \times \frac{1}{9} = \frac{8}{81}$.
(ii) p(both < 5 min) = $\frac{8}{9} \times \frac{8}{9} = \frac{64}{81}$
(iii) Either we will both check out in less than 5 minutes or at least one of us (me, my wife, or both) will be kept waiting. So, p(at least one $\geqslant$ 5 min) = 1 − p(both < 5 min)
$$= 1 - \frac{64}{81}$$
$$= \frac{17}{81}$$

Review

We are now about at the mid-point of our work on probability. Let's pause here to recall just a few of the main ideas.
(i) Fill in the missing parts of the two formulae shown below:
$$p = \frac{\text{No. of outcomes giving looked-for result}}{\text{Total no. of \_\_?\_\_ \_\_?\_\_ outcomes}}$$

$$p = \frac{\text{No. of trials giving looked-for result}}{\underline{?}}$$

(ii) Which of the two formulae is used for theoretical probabilities and which for practical probabilities?

(i) The missing words are: in the first formula, 'equally likely'; in the second, 'Total no. of trials'.
(ii) The first formula is for theoretical probability; the second for practical probability.

Fill in the blanks in the following items:
(i) Suppose you have to find the probability that either one result *or* another will happen: provided they are __?__ results, you can __?__ their separate probabilities together and so find their (*total/joint?*) probability.
(ii) Suppose you have to find the probability that a number of separate results *all* happen: so long as the results (*are/are not?*) mutually exclusive you can __?__ their separate probabilities together and so find their (*total/joint?*) probability.
(iii) To find the probability that *at least one* of a number of independent results will take place, you can either:
    (a) Subtract from. . . . . (*finish in your own words*); or
    ◦(b) add together . . . . (*finish in your own words*).

Here are the missing words:
(i)   *mutually exclusive . . . add . . . total*
(ii) *are not . . . multiply . . . joint*
(iii) (a) *subtract from 1 the probability that none of the results will take place.*
      (b) *add together the probabilities that one, two, three, etc. of the results will take place.*

Having got this far, you should be able to cope with a wide variety of probability problems. The calculations themselves are not usually difficult. The difficulty often lies in not knowing what calculation you're supposed to be doing. As I've said before, it's essential to keep a cool head and get quite clear about what kind of problem you are dealing with. As you'll remember (probability trees, Ken's buses, etc.) it's often very worthwhile to draw *diagrams* to help you sort out a probability situation.

   Professors, in particular, can seem very ingenious in providing fancy 'wrappings' for what are basically quite straightforward probability calculations. So, let's end this part of the book with some examination questions for you to try. Some are more difficult than others, but none should be beyond you if you give it enough thought. The answers will follow directly after the questions. I will not give you the working this time—I think it is a better use of my space to give you many questions without working than a few questions with the working. If in some cases you just can't see how my answer was arrived at, I suggest you get together with a teacher or with a colleague who is also studying probability.

Examination Questions

1 The probability that Ann will go to a University is $\frac{1}{5}$; the probability that she will go to a technical school is $\frac{1}{3}$. Independently the probability that Beryl will go to a University is $\frac{1}{4}$.

Calculate the probability that
(i) Ann and Beryl will both go to a University,
(ii) Ann will go to neither a University nor a technical school,
(iii) neither Ann nor Beryl will go to a University,
(iv) either Ann or Beryl (but not both) will go to a University.

2 A firm appoints a new maintenance engineer to look after three machines A, B and C. The probability that on a particular day machine A will break down is 0.2, for machine B it is 0.3 and for machine C it is 0.1. Assuming that the appointment of the new engineer makes no difference to these probabilities, what is the probability that on his first day at work
(i) all three machines break down,
(ii) exactly one machine breaks down?

3 (a) In a certain quiz game, each of three contestants can choose to answer one of three categories of question. Assuming that the contestants choose independently and that each is equally likely to select any of the categories, find the probability that: (i) all will choose the same category; (ii) all will choose different categories; (iii) two will be alike and the third different.

(b) From a pack of 52 cards, seven are taken at random, examined, and replaced. The cards are shuffled and another seven are drawn at random. Find the probability that at least one card will be drawn twice.

4 A bag contains three coins, two of them fair and the other double-headed. A coin is selected at random from the bag and tossed. If a head appears the same coin is tossed again; if a tail appears then another coin is selected at random from the two remaining coins in the bag and tossed.

By labelling the coins A, B, C and using a tree diagram or otherwise, find the probability that
(i) a head appears on the first toss,
(ii) a head appears on both tosses,
(iii) a tail appears on both tosses.

5 A bus stop is served by express buses, which run every hour at ten minutes past the hour, and local buses, which run every quarter of an hour, starting at five past the hour. What is the probability that a passenger arriving at the stop at random will have to wait more than six minutes for a bus?

6   Two players A and B play a series of independent games. For each game, the probability that A wins is $\frac{1}{4}$, the probability that B wins is $\frac{1}{5}$ and, if neither player wins, the game is considered drawn.

For a single game, calculate

(i)   the probability that the game is drawn;

(ii)   the probability that the game is either drawn or won by A.

If two games are played, what is the probability that both are won by B?

7   For a certain card game, a normal pack of 52 cards is taken and all the twos, threes and fours are removed, leaving a reduced pack of 40 cards. An Ace scores 11, the face cards (King, Queen and Jack) score 10 while the other cards score their face value. Each player receives two cards, the total score being the sum of the scores of the two cards.

In both (a) and (b) [below], answers to parts (ii), (iii) and (iv) may be given either as fractions in their lowest terms or as decimals correct to two decimal places.

(a)   A player is dealt two cards, a heart and a spade, from this reduced pack. Produce a table showing the total scores of all the possible pairs of cards which the player could have received and hence find

(i)   the most probable score,

(ii)   the probability that the player has two picture cards,

(iii)   the probability that his score is exactly 15,

(iv)   the probability that his score is 15 or less.

(b)   If *instead* the player received two hearts, use the appropriate portion of your table to find

(i)   the most probable score,

(ii)   the probability that the player has two face cards,

(iii)   the probability that his score is exactly 15,

(iv)   the probability that his score is 15 or less.

8   When a certain coin is tossed, the probability of obtaining a head is 0.4. For this coin, calculate:

(i)   the probability of obtaining a tail when the coin is tossed once;

(ii)   the probability of obtaining at least one tail when the coin is tossed twice;

(iii)   the expected number of tails obtained when the coin is tossed 50 times.

9   Box A contains 5 red, 6 white and 9 blue balls. Box B contains 7 red, 8 white and 4 blue balls. A ball is withdrawn at random from box A and placed in box B. A ball is then taken at random from Box B. Find the probability that

(i)   both balls withdrawn were red.

(ii)   the ball withdrawn from box B was red,

(iii)   the ball withdrawn from box B was either red or white.

10   Three men, A, B and C, share an office with a single telephone. Calls come in at random in the proportions $\frac{2}{5}$ for A, $\frac{2}{5}$ for B, $\frac{1}{5}$ for C. Their work requires the men to leave the office at random times, so that **A** is out for half his working time and **B** and **C** each for a quarter of theirs.

For calls arriving in working hours, find the probabilities that:

(i) no-one is in to answer the telephone;

(ii) a call can be answered by the person being called;

(iii) three successive calls are for the same man;

(iv) three successive calls are for different men;

(v) a caller who wants **B** has to try more than three times to get him.

11 In a shop, apples are sold in trays containing four apples. A random sample of 50 trays was examined for bruised apples and the results were as follows:

Number of bruised apples per tray	0	1	2	3	4
Number of trays	24	14	10	2	0

Using this table, calculate

(a) the total number of bruised apples;

(b) the probability that an apple selected at random from the 200 apples will be bruised;

(c) the probability of selecting at random from all 50 trays a tray containing exactly two bruised apples;

(d) in how many of 1000 trays you would expect to find two or more bruised apples, if it were assumed that the sample was representative of all trays sold.

12 An ordinary die has six faces numbered from 1 to 6. If an ordinary die is thrown four times, find the probability of obtaining

(a) the same number each time,

(b) four different numbers,

(c) two even numbers and two odd numbers.

Find the average score from a single throw and hence find the expected total score from four throws. (Answers may be given as fractions in their lowest terms.)

13 A multinational company recruits each year a large number of graduates. In the first two years the graduates spend a short period in a number of main departments of the company and then an assessment is made of their potential. 10% are sent on a full-time course in Management Studies for an MA Degree. 30% are allowed time off to take a Diploma in Management Studies course. The remainder are not given time off but are refunded any expenses they incur on any relevant course which they attend.

An analysis of the subsequent promotion of graduate recruits shows that 60% of those with MA in Management Studies have reached the rank of senior executive by the age of 35. The corresponding percentages for those who have had time off and no time off were 20% and 5% respectively.

(a) Find the probability that a graduate recruit randomly selected will be a senior executive by the age of 35.

(b) Find the probability that a man who is a senior executive at the age of 35 has an MA in Management Studies from a full-time course.

14 For a certain program, students may study one, two or three optional subjects. Sixteen study sociology only, 21 study politics only, 20 study history only, 8 study history and politics, 7 study sociology and politics, 5 study history and sociology, and 3 study history, politics and sociology.

How many students are there in the program? Find the probability that a student selected at random studies one optional subject only.

15 On a small island, 2 rabbits are caught. Each is tagged so that it can be recognised again and is then released. Next day 5 more (untagged) rabbits are caught. These too are tagged and released. On the third day, 4 rabbits are caught and two of them are found to be already tagged.

Assuming the rabbit population is constant and that tagged and untagged rabbits are equally likely to be caught, find

(a) the smallest number of rabbits there can be on this island,

(b) the probability of the second day's catch if there are exactly 12 rabbits on the island.

(c) the joint probability of obtaining both the second and third catches if there are exactly 16 rabbits on the island.

Answers to Examination Questions

1 (i) $\frac{1}{20}$ (ii) $\frac{8}{15}$ (iii) $\frac{3}{5}$ (iv) $\frac{7}{20}$

2 (i) $\frac{6}{1000} = 0.006$ (ii) $\frac{398}{1000} = 0.398$

3 (a) (i) $\frac{1}{9}$ (ii) $\frac{2}{9}$ (iii) $\frac{2}{3}$ (b) 0.661

4 (i) $\frac{2}{3}$ (ii) $\frac{3}{8}$ (iii) $\frac{1}{8}$

5 $\frac{31}{60}$

6 (i) $\frac{3}{5}$ (ii) $\frac{17}{20}; \frac{1}{25}$

7 (a) (i) 20 (ii) $\frac{9}{100}$ (iii) $\frac{3}{25}$ (iv) $\frac{27}{100}$

(b) (i) 20 (ii) $\frac{1}{15}$ (iii) $\frac{2}{15}$ (iv) $\frac{4}{15}$

8 (i) 0.6 (ii) 0.84 (iii) 30

9 (i) 0.10 (ii) 0.36 (iii) 0.78

10 (i) $\dfrac{1}{32}$ (ii) $\dfrac{13}{20}$ (iii) $\dfrac{17}{125}$ (iv) $\dfrac{24}{125}$ (v) $\dfrac{1}{64}$

11 (a) 40 (b) $\dfrac{1}{5}$ (c) $\dfrac{1}{5}$ (d) 240

12 (a) $\dfrac{1}{216}$ (b) $\dfrac{5}{18}$ (c) $\dfrac{3}{8}$; $3\dfrac{1}{2}$ and 14

13 (a) $\dfrac{3}{20}$ (b) $\dfrac{2}{5}$

14 80; $\dfrac{57}{80} = 0.7125$

15 (a) 9 (b) 0.318 (c) 0.1904

3
Combinations and Permutations

A Reminder

You have already learned how to calculate probabilities by using the 'equally likely outcomes' formula,

$$\frac{Probability\ of\ result}{we\ are\ looking\ for} = \frac{\text{No. of outcomes giving looked-for result}}{\text{Total no. of equally-likely outcomes}},$$

together with the addition and multiplication rules for combining probabilities.

To refresh your memory before we go on ...

Two dice are thrown together. What is the probability that the total score will be 7? If you throw the two dice again, what is the probability you'll get 7 *both* times? What is the probability you'll get at least one 7 from the two throws?

On a throw of two dice, there are 36 equally likely outcomes. Only six of these outcomes will give the looked for result (a total of 7)—that is:

4 on the first die and 3 on the second, or vice versa;

5 on the first die and 2 on the second, or vice versa

and 6 on the first die and 1 on the second, or vice versa.

So the probability of throwing a 7 is: $\frac{6}{36} = \frac{1}{6}$.

The probability of throwing a 7 on both double throws is (by the multiplication rule): $\frac{1}{6} \times \frac{1}{6} = \frac{1}{36}$.

The probability of throwing at least one 7 is equal to the difference between 1 and the probability of throwing no 7s—(since you are certain to throw either no 7s or at least one). The probability of no 7s is $\frac{5}{6} \times \frac{5}{6} = \frac{25}{36}$. (Here we are using the multiplication rule to combine the probability of not getting a 7 on the first throw with the probability of not getting a 7 on the second throw.)

And the probability of getting at least one 7 is

$$1 - \frac{25}{36} = \frac{11}{36}$$

To solve these problems with the dice, you had to know how many equally likely outcomes were possible and which of them would give you the result you were looking for. Sometimes this is not easy to see. Using our present methods it could be very tedious working out some problems.

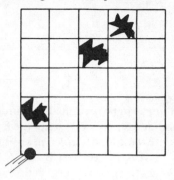

For instance, the illustration shows a window in a factory. The factory's apprentices play softball outside every lunch-time. During the course of a month, 3 of the 25 window-panes get broken by their ball. What is the probability that the broken panes all lie on one of the 14 diagonals? (One of these is illustrated.)

Admittedly, you could work out all the possible positions for the three broken panes and see how many of those lay on a diagonal, but it might take you quite a while. In the remainder of this book we'll develop easy methods for tackling problems as complex as this by using the idea of combinations and permutations.

Combining and Permuting

Suppose seven horses are running a race and you are trying to predict the winners. Which would you be more likely to predict correctly:

   (a) The *names* of the first three horses past the finishing post?

or (b) The *order* in which the first three names passed the post?

or (c) Would the two predictions be equally likely to be correct?

You would be more likely to be correct about predicting (a), the names of the first three horses. Predicting the order would be more difficult. Even if you guessed the first three names correctly, you could still be wrong about which was first, second, and third.

If we select a group of items without worrying about what order they are in (like the names of the first three horses), then we are talking about a *combination*.

But when we start considering the order or arrangement within the group (e.g. which of the names is 1st, 2nd, and 3rd) then we are dealing with a *permutation*.

Thus the names of the four horses that did not finish among the first three would make up a ___?___ , while the order in which those four did finish the race would be a ___?___ .

---

combination, . . . permutation

Any combination of things can be permuted—re-ordered or re-arranged—in several different ways. For instance, let's select the first three letters from the alphabet. We could write these letters down in six different ways:

ABC   BCA   CAB   ACB   BAC   CBA

(i)   How many combinations do we have here?
(ii) How many permutations?

(i) There is only one combination—the first three letters of the alphabet are the same three letters, no matter what order they are written in.
(ii) There are six permutations—each different order is a new permutation.

To get a new combination we must change at least one of the items in the group—we could add a new item, remove one of the present items, or do both. But to get a new permutation from a particular group (combination), we need only re-arrange its items.

Which are more numerous, combinations or permutations?

Permutations are more numerous. From each possible combination, more than one permutation can be made by re-arranging its items.

In dealing with these two concepts, try to remember that order counts in a permutation; in a combination it doesn't. This can be difficult because in ordinary usage the term "combination" is sometimes used incorrectly. For example, a "combination lock" should really be called a permutation lock. It's not enough to get the right combination of digits—you have to permute them into the right order before the lock will open.

Which of the following everyday activities involve permutations, and which involve combinations only?

(i)   Choosing a baseball team from a class of 34 boys.
(ii) Deciding which of the boys will bat first, second, and so on.
(iii) Deciding which three books to borrow from a library.
(iv) Planning the best sequence in which to decorate the rooms in my house.
(v)   Planting a row of rose-bushes so that red and yellow flowers alternate.

(i) and (iii) are combinations; (ii), (iv) and (v) are permutations.

How Many Ways?

This is a question we'll often need to ask—whether we are considering the number of ways of selecting a group of items (a combination) or the number of ways of re-arranging them (a permutation). How do we calculate this number of ways?

Let's look at some simple examples. With each one, ask yourself 'Is this a permutation problem or is it a combination problem?'

I want to travel from Boston to Los Angeles via New York. I can go by plane or boat to New York, and by plane, train or bus to Los Angeles. How many ways can I get from Boston to Los Angeles?

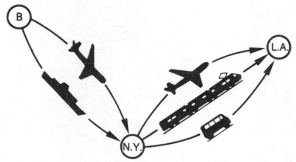

Now there are two different ways I can get to New York and then, whichever I choose, I have a choice of three ways of going on to Los Angeles. So either of the first two ways can be followed by any of the second three ways. How many ways for the overall journey?

There are six ways of doing the journey. This shows up very clearly in a tree diagram:

1st trip	2nd trip	Way of doing the journey
plane	plane	plane, plane
	train	plane, train
	bus	plane, bus
boat	plane	boat, plane
	train	boat, train
	bus	boat, bus

Here is a menu offering you a three course meal. As you see, you have a choice of 3 first courses, 4 second courses, and 2 third courses.

How many ways are there of having a meal from this menu?

As you will see from the probability tree below, there are twenty-four ways of having a meal:

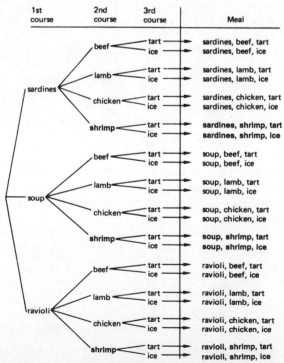

However, it isn't always easy to list all the possible combinations. And if you had, say, half a dozen choices at each step, it wouldn't even be easy to draw a tree diagram.

But perhaps you've already spotted a short-cut. Try this problem.

A firm is about to take on a new sales director and a new receptionist. There are 6 applicants for the sales director's job and 10 for that of the receptionist. How many ways can the pair be chosen?

The two new employees can be chosen in 60 ways. Any of the six applicants for the sales director's job could be linked with any of the ten possible receptionists. This gives $6 \times 10 = 60$ possible combinations.

If the company then went on to appoint a new accountant from among 4 candidates, each of those four could be linked with any of the previous 60 pairs, giving $6 \times 10 \times 4 = 240$ possible combinations.

In general terms we can say that:

> If there are a ways of doing one thing, b ways of doing a second thing, c ways of doing a third (and so on), then the number of ways of doing all these things is $a \times b \times c$ (and so on).

How many combinations of digit/consonant/vowel (in that order can be made from the phrase '1984 Olympics'? (e.g. 9yo, 4pi, etc.)

There are 4 digits, 6 consonants, and 2 vowels. Each of the 4 could be combined with any of the 6, which in turn could be combined with any of the 2. In short, this gives $4 \times 6 \times 2 = 48$ combinations. (Write out all the combinations and count them if you're not convinced.)

The total 'number of ways' for a series of choices or operations is always equal to $a \times b \times c \ldots$, where a, b, c etc. are the number of ways each operation might be done separately. This is equally true if the number of ways is the same for each of the operations—that is, if $a = b = c$ etc.

For instance:

There are four possible routes from police HQ to the bus station, another four from the bus station to the bank, and four more from the bank to police HQ. How many ways can a police car patrol a circuit from police HQ to bus station to bank and back to HQ?

The patrol car can deal with the trip to the bus station in any of 4 ways; once it has done that it can deal with the trip from bus station to bank in any of another 4 ways; and then the trip back to HQ can also be done in any of 4 ways. Hence, there are $4 \times 4 \times 4 = 64$ combined routes.

If you are making an oil and vinegar dressing, you can make it plain or you can add some combination of salt, pepper, mustard, and garlic (or all of them!). So how many different dressings can you make? (If you need a hint, look at the footnote*)

We deal with the extra ingredients one at a time. We can treat the salt in either of two ways—add it or leave it out. Once we've dealt with the salt, we can treat each of the other ingredients in the same two ways. Thus we can make $2 \times 2 \times 2 \times 2 = 2^4 = 16$ possible dressings (including the plain one and the one with everything).

Here is a similar problem where each 'item' can be dealt with in the same number of ways.

I have three cousins. In how many ways can I invite one or more of my cousins to a party? (Remember to omit the case where I don't invite any of them.)

I can deal with my first cousin in either of two ways—by inviting him or not. After I have dealt with him, I can deal with my second cousin in the same two ways. Similarly with my third cousin. This gives $2 \times 2 \times 2 = 8$ different results. However, in one of these cases, none of my cousins was invited. So this gives 7 different parties to which one or more of my cousins would be invited. (Check this out with a tree diagram if you're not convinced.)

One more example of this kind:

A hospital uses a light-signal system for letting its doctors know when they are needed somewhere. Each set of lights has four colours which may be lit up four, three, two, or one at a time. On seeing his or her particular combination of colours, the doctor is expected to report to a control office.

How many doctors could each have his or her own combination of lights?

Each of the four lights could be dealt with in two ways—each could be on or off. So the total number of ways the signals may be set is

---

* Deal with the four extra possible ingredients one at a time. Each can be dealt with in either of two ways—add it or leave it out.

$2 \times 2 \times 2 \times 2 = 16$. But this includes the case where all the lights are off. That would be no use as a signal, so we are left with 15 distinct combinations. Thus 15 doctors could each be given a personal colour combination.

Unfortunately, the hospital has more than 15 doctors. So it decides that each of the fifteen patterns of lights can appear in two different ways—steady or flickering. How many doctors can the system cope with now?

Each of the fifteen patterns can be dealt with in either of two ways. This allows the system to signal $15 \times 2 = 30$ different doctors.

And now a probability question to remind you of what this book is supposed to be about.

A local snack bar serves four kinds of soup, and sandwiches that can have eleven different fillings and be on white bread or wheat. You go to the snack bar one day and order a bowl of soup and a sandwich. I do the same thing next day. What is the probability that I make exactly the same choice as you? (If you need a hint, read on: How many ways can I choose soup and a sandwich? Only one of these will be the same as your choice. So I have a one in (*how many?*) chance of making the same choice as you.

With 4 kinds of soup, 11 kinds of sandwich filling, and a choice of 2 kinds of bread, there are $4 \times 11 \times 2 = 88$ ways you can choose soup and a sandwich. Therefore, I have one chance in 88 of making the same choice as you.

Now think a moment about the problems we have worked with so far. You have considered the number of ways of selecting:

   ways of traveling to Los Angeles
   courses in a meal
   a new pair of employees
   routes for a police car
   ingredients for oil and vinegar dressing
   light patterns for a signal system.

What, then, have you been dealing with so far? Combinations or permutations or both?

In none of the examples have you had to consider the order of the items you chose. (To have had your ice cream first, and then your soup, and then your chicken, for example, would *not* have counted as another 'way' of having a meal!) You have been dealing, so far, with combinations only.

Now let's go on to try a problem in permutations. Which of the following two problems involves permutations, where different orders count as extra 'ways'?
   (a) 'On getting off a train I find I have a quarter, dime, and nickel in my pocket. How many ways can I give the porter a tip?'
or (b) 'In how many different ways can three people line up at a bus-stop?'

The problem in permutations is (b). You have the combination or collection of people given you. The problem you are faced with is one of *order*—in how many ways can you arrange them? In the other problem, order was not involved: the issue was simply one of how many different combinations of coins could have been given.

Filling Places

One method for tackling the problem above (and others like it) is to say that we have three places in the line to fill—one for each person.

In the first place we can put person A, or B, or C. So that place can be dealt with in three ways.

Once we have filled this first place (with either A, B, or C) in how many different ways can we fill the second place?

Having put one of the three people in the first place, we have a choice of two only for the second place.

So we've chosen one of the three passengers for our first place and one of the remaining two for our second place.

How many ways can we fill the final place?

There is only one way of filling the final place. (With the one passenger remaining who has not yet been given a place.)

So we have 3 ways of dealing with the first place, 2 ways of dealing with the second, and 1 way only of dealing with the 3rd.

How many ways altogether are there of *forming a line* from these three people? (Draw a tree diagram, if you're not sure.)

6 different lines could be formed. There are 3 ways of dealing with the first place; then, when that has been done, 2 ways of dealing with the second place; leaving 1 way of dealing with the 3rd place. So there are $3 \times 2 \times 1 = 6$ ways of dealing with the series of choices. We can illustrate this with a tree diagram:

1st place	2nd place	3rd place	Possible arrangements

```
        B ——————— C        ABC
    A <
        C ——————— B        ACB

        A ——————— C        BAC
    B <
        C ——————— A        BCA

        A ——————— B        CAB
    C <
        B ——————— A        CBA
```

Note that the same three passengers appear in each possible arrangement. Only the order is different.

Now try another:

Five children are running a race. How many different ways could they finish the race? (That is, in how many different orders could they cross the finishing line?)

The race could finish in 120 different ways.

If we think in terms of five places to be filled, the first place can be filled by any of the five children. Once that has been done, the second place can be filled by any of the remaining four children, and so on until there is only one child left to fill the fifth place:

$$\boxed{5} \boxed{4} \boxed{3} \boxed{2} \boxed{1}$$

And $5 \times 4 \times 3 \times 2 \times 1 = 120$. So the five children could finish the race in any of 120 different orders.

But suppose we have prizes for the *first three* children only. How many ways can the first three places be filled?

The first three places can be filled in 60 different ways. With three places, the first can be filled by any of the five children; the second by any of the remaining four; and the third by any of the other three children:

$$\boxed{5} \boxed{4} \boxed{3}$$

And $5 \times 4 \times 3 = 60$.

A group of 20 people are in a meeting and wish to elect a chairman and a secretary. In a moment I'll ask you how many ways they can do this. But first, how many 'places' do we have to fill?

Since we need a chairman and a secretary there are 2 places to be filled. And we have 20 people to choose from.

How many ways can we select a chairman and a secretary?

Any of the 20 people can be chosen as chairman. But once he/she has been chosen, there are only 19 left from whom to choose a secretary. Any of the 20 could be linked with any of the 19, giving $20 \times 19 = 380$ possible pairs (possible ways of choosing a chairman and a secretary).

Mr and Mrs Thirkettle were two of the 20 people at the meeting. How many of those 380 possible pairs will include *both* Mr and Mrs Thirkettle?

Mr and Mrs Thirkettle will be included in two of the 380 possible pairs. In one, Mr Thirkettle would be the Chairman and Mrs Thirkettle the Secretary; in the other it would be the opposite way round. The one combination (Mr and Mrs) gives two permutations.

Now try this one.
The Witherspoon family (Dad, Mom, Sue, and Joe) move to a new district, and they decide that each person will register with a different dentist. There are 6 dentists in the district, and we'll need to know how many ways this family of 4 can register. But first: how many 'places' do we have to fill?

There are four places to fill. We need a different dentist for each member of the Witherspoon family, and we have 6 dentists to choose from.

How many different ways can this family register?

The first person to choose (e.g. Mom) can have any of the 6 dentists. This leaves 5 for the second person (e.g. Dad) to choose from. The third person (say Sue) can then choose from 4, leaving any of the remaining 3 for Joe.

Mom	Dad	Sue	Joe
6	5	4	3

Thus there are $6 \times 5 \times 4 \times 3 = 360$ ways of choosing and permuting 4 dentists from a group of 6 dentists. As you can confirm for yourself, it wouldn't matter which member of the family was given first, second, third and fourth choice. The same number of orders would arise.

Sometimes we'll find permutations where each 'place' can be filled in the *same* number of ways.

A number lock (wrongly called a combination lock) has four rings, each with ten different digits (0 to 9). How many different attempts to open the lock might one need to make in order to find the correct permutation?

One might make 10 000 different attempts. We have four 'places' to fill (one for each ring) and 10 digits to choose from each time:

$\rightarrow 10^4 = 10\,000$

In this case, we would have a probability of $\dfrac{1}{10\,000}$ of choosing the correct permutation first time.

How many 3-letter code words (permutations) could you make from the word STRYZI if
  (i) you can repeat letters?
  (ii) You cannot repeat letters?

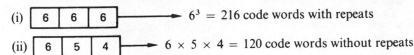

(i) | 6 | 6 | 6 | $\longrightarrow$ $6^3 = 216$ code words with repeats

(ii) | 6 | 5 | 4 | $\longrightarrow$ $6 \times 5 \times 4 = 120$ code words without repeats

How many car registration plates could you make up, using two letters followed by four digits (e.g. AD1066)? (Zero is not allowed as the first of the four digits.)

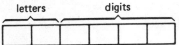

Don't do the multiplication. Just set out your figures in the six places.

You should have | 26 | 26 | 9 | 10 | 10 | 10 |

You have a choice of 26 letters each time for the first two places. For the first digit, you have 9 choices only, because zero is not allowed. Thereafter, any of 10 digits can be chosen for each of the other places. The total number of permutations is: $26 \times 26 \times 9 \times 10 \times 10 \times 10 = 6\,084\,000$.

Until the 1930s, British car registration plates had two letters followed by four digits, e.g. AX2989. Then this was changed to three letters followed by three digits (e.g. AXY298).

Can you see any advantage in the new system? How would it affect the total number of different registration plates that could be issued?)

Until the 1930s, only $26 \times 26 \times 9 \times 10 \times 10 \times 10 = 6\,048\,000$ different permutations were available. But, with the first *three* places occupied by letters, the number available increased to $26 \times 26 \times 26 \times 9 \times 10 \times 10 = 15\,818\,400$—more than $2\frac{1}{2}$ times as many.

In 1963 the system was modified again. Another letter was added *after* the digits (e.g. AXY298K). This meant that, in theory anyway, each letter of the alphabet could be linked with each of the previous registrations, giving 26 times as many possible number plates as before 1963. In fact, many of the letters are not being used, e.g. the letter O because it looks like the digit zero, the letter I because it looks too much like 1, and so on.

Next, let's consider a rather tricky point. Suppose I asked you how many 4-digit numbers you could form from the digits 3, 4, 5, 6, 7—if repeats are *not* allowed. You would know that you had four places to fill, and you could do this in $5 \times 4 \times 3 \times 2 = 120$ ways.

Now suppose instead that I ask you how many 4-digit numbers can be formed from 3, 4, 5, 6, 7, without repeats, if the numbers must all be *odd*.

Clearly, there are still four places to fill. But the last place must be filled with one of three digits. What are they?

The last place must be filled with either 3 or 5 or 7. Otherwise, the resulting number will not be odd.

So, we have three ways of filling the last place if we are to get an odd 4-digit number from the five digits 3, 4, 5, 6, and 7.

1st	2nd	3rd	4th
			3

After we have filled the last place in any of 3 ways, we can fill the first place in any of . . .

(a) 2 ways?   or (b) 3 ways?   or (c) 4 ways?

You have used one of the five digits to fill the last place, so any of the remaining four can be used in the first place, i.e. there are 4 ways of filling the first place (without repeats).

So, any of the odd digits can fill the last place. Then any of the even digits, or any of the odd digits not used in the last place, can be put in the first place—4 possibilities in all. Once this has been done, there are 3 digits left for the second place; and this leaves just 2 ways of filling the third place.

4	3	2	3

So, from the digits 3, 4, 5, 6, and 7, how many odd, 4-digit numbers (without repeats) can we form?

We can form $4 \times 3 \times 2 \times 3 = 72$ numbers. Notice: in this case we filled the fourth place first because it had to be done in a *special* way—a way that would affect the number of choices we would have for each of the other places.

Normally, it doesn't make any difference which place you fill first—you'll still end up with the same permutations. If you are, for instance, putting three books on a shelf, you can fill the middle place or either of the two end-places first: but you would still get $3 \times 2 \times 1$ or $2 \times 3 \times 1$ or $2 \times 1 \times 3$ etc permutations. Even if you put all three books on the shelf *at once*, it is still a help, when looking at such a problem, to think of it as one operation after another. But if one of the places has to be filled in a *special* way, it is usually necessary to do it *first*.

Apply this rule to the next problem.

Five friends are setting out on a journey in a 5-seater car. In how many ways can they seat themselves for the journey if only two of them can drive?

With five places (seats) to fill, one of them (the driver's) can be filled in only two ways—so we deal with it first:

driver → | 2 | 4 | 3 | 2 | 1 |

The remaining places can be filled in 4 ways, 3 ways, 2 ways, and 1 way respectively. This gives $2 \times 4 \times 3 \times 2 \times 1 = 48$ different possible seating arrangements.

Here is a slightly tougher example of the same idea: A pop-group's 'spot' on a variety show is to consist of three vocal numbers and two instrumentals. In how many ways can the 'spot' be arranged so that it begins and ends with a vocal, and neither instrumental follows directly after the other?

The order must be as follows:

Vocal   Instrumental   Vocal   Instrumental   Vocal

Since the group has 3 vocals to choose from, the first place can be filled in any of 3 ways, the third place can be filled in any of 2 ways, and 1 vocal is left for the fifth place:

| 3 | | 2 | | 1 |

With 2 instrumental numbers to choose from, the group can deal with the second place in either of 2 ways, leaving just 1 way of filling the fourth place:

| 3 | 2 | 2 | 1 | 1 |

And $3 \times 2 \times 2 \times 1 \times 1 = 12$ possible arrangements.

Restricted Neighbours

Now, to finish this section, let's look at another kind of restriction that could be put upon the permutations we are allowed to make. Consider the word NUMBER. You'll easily see how many 6-letter code-words you could make up from this word. But suppose I asked you only for those permutations in which the two vowels (U and E) are next to each other? Would you be able to tell me (without listing all the possibilities and counting them)?

First, how many 6-letter permutations can you make from the word NUMBER, paying *no* attention to where the vowels are?

We could make $6 \times 5 \times 4 \times 3 \times 2 \times 1 = 720$ permutations of 6 letters from the word NUMBER.

But in how many of these 720 permutations will the two vowels (U and E) be next to each other?

The best way of tackling this is to regard the two vowels as a single unit, UE (or EU). This now gives us five items to permute: N, UE, M, B, and R. (This is like treating U and E as a single letter.) This gives us $5 \times 4 \times 3 \times 2 \times 1 = 120$ permutations.

So, out of the 720 permutations of the letters in the word NUMBER, 120 will have the vowels next to each other in the order UE. However, the vowels would still be next to each other if the order were reversed—EU instead of UE.

So how many permutations will there be altogether, from the word NUMBER, in which the two vowels are next to each other?

Altogether there will be $2 \times 120 = 240$ permutations in which the letters U and E are together (either UE or EU). If you doubt this, try listing them; I think you'll soon get tired.

Note:
We started with 6 letters.
But two of them had to be side by side in permutations.
So we treated those two as one letter.
This left 5 letters, giving $5 \times 4 \times 3 \times 2 \times 1$ permutations.
Each of these permutations has the two letters together in a certain order.
They are also together (but in reverse order) in another
$5 \times 4 \times 3 \times 2 \times 1$ permutations.
Thus, they are together (in one or other order) in
$2(5 \times 4 \times 3 \times 2 \times 1)$ permutations.

Without restrictions, we decided that the letters in NUMBER give 720 permutations. In how many of these are the vowels *not* side by side?

Since we know that the vowels are together in 240 of the 720 permutations, then in the remaining $720 - 240 = 480$ they must be apart.

Try another one.
In how many ways can seven anglers be lined up along a beach in a fishing contest if the two champion anglers are to be ...
  (i) put next to each other?
  (ii) kept apart?

(i) To calculate the number of permutations in which the two champion anglers are together, we must treat them as one. This leaves us with six anglers, giving $6 \times 5 \times 4 \times 3 \times 2 \times 1 = 720$ permutations in which the two champions are together. However, their order can be changed *within* those permutations, giving twice as many: $2 \times 720 = 1440$.
(ii) And to find the number of permutations in which the champions would be apart, we must subtract 1440 from the total number of permutations that would be available with all seven anglers, i.e. $7 \times 6 \times 5 \times 4 \times 3 \times 2 \times 1$

= 5040. And 5040 − 1440 = 3600 permutations in which the two champions will be kept apart.

Suppose the seven anglers were assigned their positions by drawing names at random from a hat. What is the probability that the two champion anglers would be *apart* from one another? (If you need a hint, look at the footnote*.)

The random drawing of names could give any of
$7 \times 6 \times 5 \times 4 \times 3 \times 2 \times 1 = 5040$ permutations. Of these,
$2(6 \times 5 \times 4 \times 3 \times 2 \times 1) = 1440$ would put the two champions side by side; thus 5040 − 1440 = 3600 would keep them apart. So they have 3600 chances in 5040 of being separated—a probability of $p = \dfrac{3600}{5040} = \dfrac{5}{7}$.

Finally, try this question:
Six tulip bulbs, of which only two produce yellow flowers, are planted in a row. What is the probability that the yellow tulips will be next to each other?

Altogether, the six flowers can be permuted in
$6 \times 5 \times 4 \times 3 \times 2 \times 1 = 720$ ways.

   If two of the flowers are to be side by side, we'll count them as one: this gives $5 \times 4 \times 3 \times 2 \times 1 = 120$ ways of arranging the flowers with two of them next to each other in a particular order. If we remember that the two yellow flowers will still be next to each other if we swap them around, we see that we have $2 \times 120 = 240$ permutations.

   Since there are 240 ways of getting the two yellows together, out of a total of 720 ways of arranging all six flowers, the probability of getting the two yellow flowers next to each other is $\dfrac{240}{720} = \dfrac{1}{3}$.

In this last section, you've met quite a few problems involving combinations or permutations. The arithmetic is usually child's play; but, as with all questions of probability, it is essential to think very clearly and to reason around a problem until you see how to tackle it.

   Before going on to the next section, here are some review questions for you to try.

Review Questions

1   A boy has five pairs of trousers and six shirts. His girl-friend has four skirts and seven tops. Who has the most choice of what to wear?

2   A shop sells eight kinds of bread, six kinds of butter, and five kinds of

*Simply ask yourself these questions: (a) how many different arrangements are possible? and (b) how many of these arrangements would keep the champions apart? And use your answers to calculate the probability fraction.

cheese. If two women go to the shop independently and each buys bread, cheese and butter, what is the probability that both have chosen the same kinds?

3   In a newspaper competition, 12 reasons are suggested for preferring Brand X bicycle oil. To enter the competition you must select the five 'best' reasons and arrange them in order of importance. If each entry costs $2, how much would it cost you to send in every possible entry?

4   Four prisoners are being moved by train from one jail to another. If the train has eight coaches, and each prisoner must travel in a different coach, in how many possible ways can the men travel on the train? How many ways would there be if they did not have to travel in different coaches?

5   Which of the above questions deal with combinations, and which with permutations?

6   Hilda and Mabel, who know nothing about horses, go to a race meeting. Hilda tries to guess the first three places in a race with 5 horses, while Mabel tries to guess the first five places in a race with 7 horses. Who has the greater chance of being right?

7   Certain registration numbers are formed from three different letters, chosen from the first ten letters of the alphabet, followed by a three figure number which must not begin with zero. Calculate how many registration numbers can be formed.

8   In how many ways can 6 books be arranged on a shelf if (a) two particular books must be side by side, and (b) if these two books must not be side by side?

9   A goods train is being made up of 3 cattle trucks, 2 ore carriers, and 4 coal wagons. In how many ways can the train be assembled if the 2 ore carriers must come at the front and the 4 coal wagons must bring up the rear? (The order within each group matters.)

10   The police are wanting to interview a man known to drive a car with a seven-unit registration, beginning with either P, B, or D, followed by either AR69 or KR69, then an unknown digit between 0 and 9, and finally a letter before G in the alphabet. If it takes 15 minutes to check the owner of each possible registration with the licensing office, how long will it take to name the owners of all registrations fitting the known facts? Only after this donkey-work is complete is it reported that the final letter of the registration was F. How much time would have been saved if this had been known in the first place?

11   In how many ways can 5 unlike objects be arranged in order? A party of 3 ladies and 2 gentlemen goes to a theatre and sits in a row of 5 seats in a random order. Find the probability that
   (a) the two gentlemen sit together,
   (b) the ladies and gentlemen occupy alternate seats.

12   Which of the questions 6–11 deal with combinations, and which with permutations?

Answers to Review Questions

1 Boy has $5 \times 6 = 30$ choices; girl has $4 \times 7 = 28$ choices.

2 The first woman has $8 \times 6 \times 5 = 240$ ways of choosing her bread, butter and cheese. The second woman's chance of matching this selection is, therefore, $\frac{1}{240}$.

3 5 reasons can be chosen from 12 and arranged in $12 \times 11 \times 10 \times 9 \times 8 = 95040$ ways. At \$2 an entry this means $95040 \times \$2 = \$190080$.

4 There are $8 \times 7 \times 6 \times 5 = 1680$ ways for the four prisoners each to travel on a different coach of the 8-coach train. But if they don't have to travel in different coaches there are $8 \times 8 \times 8 \times 8 = 4096$ ways for them to travel.

5 Questions 1 and 2 deal with combinations; questions 3 and 4 go on to permutations.

6 Chance of success: Hilda: $\dfrac{1}{15 \times 14 \times 13} = \dfrac{1}{2730}$

Mabel: $\dfrac{1}{7 \times 6 \times 5 \times 4 \times 3} = \dfrac{1}{2520}$

7 There are ten choices for the first letter, 9 for the second, and eight for the third. Since the number cannot begin with 0, we have nine choices for the first digit, but ten for the second and for the third. So we can form $10 \times 9 \times 8 \times 9 \times 10 \times 10 = 648\,000$ registration numbers.

8 Six books can be arranged on a shelf . . .

(a) with two particular books side by side in $2(5 \times 4 \times 3 \times 2 \times 1) = 120$ ways.

(b) with those books not side by side in $720 - 120 = 600$ ways. (600 is the number of ways of permuting 6 books without any restrictions at all, i.e. $6 \times 5 \times 4 \times 3 \times 2 \times 1 = 600$.)

9 There are nine 'places' to fill, and they can be filled in $2 \times 1 \times 3 \times 2 \times 1 \times 4 \times 3 \times 2 \times 1 = 288$ ways.

10 There are seven 'places' to fill: 3 ways of filling the first, 2 for the second to fifth inclusive, 10 for the sixth, and 6 ways of filling the seventh place— $3 \times 2 \times 10 \times 6 = 360$ possible registrations. At 15 minutes a check, this would take 90 hours. If the final letter had been known, only $3 \times 2 \times 10 = 60$ registrations would have needed checking; taking 15 hours only, and thus saving 75.

11 5 unlike objects can be arranged in $5 \times 4 \times 3 \times 2 \times 1 = 120$ ways.

(a) The two gentlemen will be sitting together in $2(4 \times 3 \times 2 \times 1) = 48$ of the possible arrangements. So the probability that they will be sitting together is $\dfrac{48}{120} = \dfrac{2}{5}$.

(b) The only possible 'alternate' order for 3 ladies (L) and 2 gentlemen (G) is L-G-L-G-L. There are 3 choices for the first place, 2 for the second, 2 for the third, 1 for the fourth, and 1 for the fifth: $3 \times 2 \times 2 \times 1 \times 1 = 12$ such arrangements. So the probability $= \dfrac{12}{120} = \dfrac{1}{10}$.

12   Questions 6–11 all deal with permutations.

Formulae for Permutations

So now you know how to tackle permutation problems. You think of so many 'places' to be filled in order. You decide how many ways there are of filling each place. You multiply these 'numbers of ways' together.

At this stage, it might be helpful for you to know some *shorthand* for writing down your calculations. If I asked you how many ways the 8 coaches of a train might be arranged, you could begin like this:

$$8 \times 7 \times 6 \times 5 \times 4 \times 3 \times 2 \times 1 = ?$$

But you could save yourself quite a bit of trouble (and space) by using the 'multiplication DOT' instead of the usual cross-sign. Like this: $8 \cdot 7 \cdot 6 \cdot 5 \cdot 4 \cdot 3 \cdot 2 \cdot 1 = ?$

The dot can be a much neater way of writing out multiplications— especially when these run into several factors, as they often do with permutations and combinations.

In a team of nine baseball players, how many possible batting orders are there? (Just lay out the factors, using the multiplication dot—don't multiply them.) There are $9 \cdot 8 \cdot 7 \cdot 6 \cdot 5 \cdot 4 \cdot 3 \cdot 2 \cdot 1$ possible batting orders.

Similarly, if we had 12 people to arrange in order, we could write:

$$12 \cdot 11 \cdot 10 \cdot 9 \cdot 8 \cdot 7 \cdot 6 \cdot 5 \cdot 4 \cdot 3 \cdot 2 \cdot 1$$

which, of course, means 'multiply all these numbers together'. If we had 13 people to permute, the total number of ways would be:

$$13 \cdot 12 \cdot 11 \cdot 10 \cdot 9 \cdot 8 \cdot 7 \cdot 6 \cdot 5 \cdot 4 \cdot 3 \cdot 2 \cdot 1$$

Notice that each of these multiplications starts with a certain number (11, 12, or 13 in these cases), and then multiplies through with a string of numbers, each of them one less than the one before it. The last number in the sequence is *always* 1.

Using Factorials

Such a multiplication series is called a *factorial*. Thus:

Factorial 11 = $\quad 11 \cdot 10 \cdot 9 \cdot 8 \cdot 7 \cdot 6 \cdot 5 \cdot 4 \cdot 3 \cdot 2 \cdot 1 = 39\,916\,800$

Factorial 12 = $\quad 12 \cdot 11 \cdot 10 \cdot 9 \cdot 8 \cdot 7 \cdot 6 \cdot 5 \cdot 4 \cdot 3 \cdot 2 \cdot 1 = 479\,001\,600$

Factorial 13 = $13 \cdot 12 \cdot 11 \cdot 10 \cdot 9 \cdot 8 \cdot 7 \cdot 6 \cdot 5 \cdot 4 \cdot 3 \cdot 2 \cdot 1 = 6\,227\,020\,800$

What is Factorial 5?

Factorial 5 = $5 \cdot 4 \cdot 3 \cdot 2 \cdot 1 = 120$

To write a factorial, we start off with the number whose factorial we require; then we multiply by number after number, going one less each time, until we end up with 1.

What is the value of:
(i) Factorial 7?
(ii) Factorial 6?
(iii) Factorial 3?

(i)   Factorial $7 = 7 \cdot 6 \cdot 5 \cdot 4 \cdot 3 \cdot 2 \cdot 1 = 5040$
(ii) Factorial $6 = \qquad 6 \cdot 5 \cdot 4 \cdot 3 \cdot 2 \cdot 1 = \quad 720$
(iii) Factorial $3 = \qquad\qquad\qquad 3 \cdot 2 \cdot 1 = \qquad 6$

Even using dots instead of the usual multiplication signs, factorials can take up a lot of room. Imagine how much space you would need to indicate the permutations in a pack of playing cards—factorial 52.

Fortunately, there is a shorthand way of writing a factorial. To indicate a factorial, we simply write an *exclamation mark* after the number whose factorial we are concerned with. Thus:

Factorial 7 is written   7!
Factorial 52 is writter 52!

In general, Factorial $n = n!$

I'm afraid this shorthand sign is going to make our pages look rather over-dramatic from now on. (Not surprisingly, ! is sometimes called the 'shriek-sign' !.) Nevertheless, it will save us a lot of space.

We can draw up a table to show how rapidly the factorials increase in value:

n	1	2	3	4	5	6	7	8	9	10
$n!$	1			24	120				362 880	

Complete the table yourself.

The missing factorials are as follows:
$2! = \qquad 2$
$3! = \qquad 6$
$6! = \qquad 720$
$7! = \qquad 5040$
$8! = \qquad 40\,320$
$10! = 3\,628\,800$

> Keep a copy of the completed table where you can get at it easily; you'll need to refer to it often during the rest of this book.

And after 10!, the factorials continue to increase at a fantastic rate. For instance, suppose you take a class of 30 children and try to seat them in every possible arrangement—30! altogether. If you had all the men, women and children on earth to help you, you would never manage to get through all the permutations—not if you worked day and night for more than a million years. Why?

Because 30! $\simeq$ 265 250 000 000 000 000 000 000 000 000 000

Use your completed table to check the number of arrangements you could get by permuting all the letters of the word WONDERFUL. How Many?

WONDERFUL has 9 letters; but we don't have to calculate factorial $9 = 9 \cdot 8 \cdot 7 \cdot 6 \cdot 5 \cdot 4 \cdot 3 \cdot 2 \cdot 1$ in full, because the table tells us that 9! = 362 880.

Before we go on, let's look at a couple of very useful tricks we can play with factorials.

1   We can *stretch* them:

e.g. $100! = 100 \cdot 99!$
$$52! = 52 \cdot 51 \cdot 50 \cdot 49!$$
$$7! = 7 \cdot 6! = 7 \cdot 6 \cdot 5! = 7 \cdot 6 \cdot 5 \cdot 4! \text{ etc.}$$

2   You can also *divide* factorials by cancelling:

e.g. $\dfrac{7!}{4!} = \dfrac{7 \cdot 6 \cdot 5 \cdot \cancel{4!}}{\cancel{4!}} = \dfrac{7 \cdot 6 \cdot 5 \cdot \cancel{4 \cdot 3 \cdot 2 \cdot 1}}{\cancel{4 \cdot 3 \cdot 2 \cdot 1}} = 210$

Use these two tricks to find the value of $\dfrac{6!}{3!}$

$$\frac{6!}{3!} = \frac{6 \cdot 5 \cdot 4 \cdot \cancel{3!}}{\cancel{3!}} = \frac{6 \cdot 5 \cdot 4 \cdot \cancel{3 \cdot 2 \cdot 1}}{\cancel{3 \cdot 2 \cdot 1}} = 120$$

Try a couple more.

What is the value of:

(i) $\dfrac{3!}{4!}$?

(ii) $\dfrac{6!3!}{4!2!}$? (That is factorial 6 multiplied by factorial 3, divided by factorial 4 multiplied by factorial 2.)

(i) $\dfrac{3!}{4!} = \dfrac{\cancel{3!}}{4 \cdot \cancel{3!}} = \dfrac{\cancel{3 \cdot 2 \cdot 1}}{4 \cdot \cancel{3 \cdot 2 \cdot 1}} = \dfrac{1}{4}$

(ii) $\dfrac{6!3!}{4!2!} = \dfrac{6 \cdot 5 \cdot \cancel{4!} 3 \cdot \cancel{2!}}{\cancel{4!}\cancel{2!}} = 90$

Now that you know how to handle factorials, let's see how they help with permutations. You know that if you have to permute, say, 3 different objects in every possible way, the total number of permutations is 3! Four objects would allow 4! different arrangements. In general, *n* objects give *n*! permutations.

In general terms, if you have to permute *n* objects, you will have a total of (*how many!*) permutations.

---

n! permutations.

Thus, if we were to take four different objects and permute all four of them in every possible way, we would get 4! different arrangements or permutations. Now for a bit more shorthand: instead of saying

The number of different permutations obtained by taking 4 different objects and arranging all 4 of them in every possible way equals 4!

we can write $4P4 = 4!$

Here the **P** stands for 'the number of different permutations' and the numbers before and after the **P** tell how many objects we have to choose from and how many will appear in each of the permutations (all of them, in this case).

Another example, using this shorthand:

The number of items obtained by taking 8 items and arranging all 8 in every possible way $= 8P8 = 8!$

How would you use the shorthand to represent the number of permutations obtained by arranging a class of 30 children in every possible way.

30**P**30 is the 'number of permutations obtained by taking 30 items and arranging all 30 of them in every possible way.

$$\rightarrow 30P30 = 30!$$

Similarly,

3P3	$3 \cdot 2 \cdot 1$	$= 3!$
4P4	$4 \cdot 3 \cdot 2 \cdot 1$	$= 4!$
5P5	$5 \cdot 4 \cdot 3 \cdot 2 \cdot 1$	$= 5!$
6P6	$6 \cdot 5 \cdot 4 \cdot 3 \cdot 2 \cdot 1$	$= 6!$

In general, the number of permutations of *n* objects taken all together is $nPn = n!$

Thus, if I told you that 5 candidates replied to a job advertisement, and asked you in how many 'orders of preference' they could be placed, you might say '5!', or '$5 \cdot 4 \cdot 3 \cdot 2 \cdot 1$', or you might simply say that the number of permutations is ? **P** ?

$5P5 = 5 \cdot 4 \cdot 3 \cdot 2 \cdot 1 = 5!$

Here we have taken 5 objects (job candidates) and considered every possible arrangement in which *all 5* could appear. However, as you'll remember from the last section, we sometimes take a set of objects and permute just *some* of them at a time. (We have fewer 'places' to fill than the total number of objects.)

For instance, suppose we are not really interested in the number of possible ways in which all five of our candidates could be placed. The fact is we have just *two* jobs to offer—Chief Editor and Assistant Editor. What we want to

know is: how many ways can we fill these posts? (Here we are considering all the possible permutations of 5 people taken *2 at a time*.)
    So how many permutations are there?

There are 20 possible permutations. Call the candidates A, B, C, D and E. There are 5 ways we can choose the Chief Editor. Once he or she is chosen, there are 4 candidates remaining from whom to choose the Deputy Editor. So the number of possible permutations is 5 × 4 = 20.

We could show all the possible arrangements in a tree diagram, or simply in a table like this:

Chief Editor	A	A	A	A	B	B	B	B	C	C	C	C	D	D	D	D	E	E	E	E
Deputy Editor	B	C	D	E	A	C	D	E	A	B	D	E	A	B	C	E	A	B	C	D

20 possible permutations

Here we have shown the permutations of 5 items taken 2 at a time. We can write this more simply as 5P2:

This number of items → 5P2 ← This many permuted at a time

How would you write the number of permutations of 5 objects taken 3 at a time?

5P3 is the number of permutations of 5 things taken 3 at a time.

The number *before* the **P** | → 5P3 ← | The number *after* the **P** tells how
tells how many things we | | many of these things are to be in
have to choose from | | each arrangement.
Thus, 25P11 stands for all the possible permutations of _?_ things at a time that can be made from a group totalling _?_ things in all.

Thus, 25P11 stands for all the possible permutations of *11* things at a time that can be made from a group totalling *25* things in all.

In general, then:

Say we have *n* things and we wish to form permutations each containing *r* of those things:
The *number of permutations* of *n* things taken *r* at a time is
$$n\mathbf{Pr}$$

Using this shorthand, how many ways could you arrange 10 books taking 4 of them at a time?

The number of permutations from 10 items taken 4 at a time is 10**P**4.

Now try writing the n**P**r symbol for each of these problems.
(i)   In how many ways can 3 men each be given a hotel room of his own if there are 8 free rooms available?   **? P ?**
(ii)   There are 6 routes between two towns. How many ways are there of going from one place to the other and returning by a different route?   **? P ?**
(iii)   A man can cook 9 different dinners. How many different sequences of dinner could he cook for a week if he didn't prepare the same dinner twice.

? P ?

(i)   Here we have 8 rooms to be permuted 3 at a time (among the three men): 8**P**3
(ii)   Permutations of 2 routes with 6 to choose from: 6**P**2
(iii)   9 meals to choose from, so how many permutations of 7 (one for each day of the week) at a time: 9**P**7

Perhaps we should now try to work out the actual number of permutations that some of these shorthand expressions stand for.
You'll remember we found 5**P**5 $= 5 \cdot 4 \cdot 3 \cdot 2 \cdot 1 = 120$
        and 5**P**2 $= 5 \cdot 4$ $= 20$
What is 5**P**3? (You are filling three 'places' with 5 to choose from.)

5**P**3 $= 5 \cdot 4 \cdot 3 = 60$
Here we are permuting 5 things, taking 3 at a time. With 3 'places' to fill we can fill the first in 5 ways, the second in 4 ways and the third with any of the remaining 3 things.

One way of looking at n**P**r is to say
'Beginning with n, write r factors each being
one less than the previous factor.'
Thus, if $n = 9$ and $r = 7$, n**P**$r = 9$**P**7
                $= 9 \cdot 8 \cdot 7 \cdot 6 \cdot 5 \cdot 4 \cdot 3$ (7 factors)
What is the value of:
(i)   6**P**2?
(ii)   8**P**3?
(iii)   10**P**4?

(i)   6**P**2 $= 6 \cdot 5 = 30$
(ii)   8**P**3 $= 8 \cdot 7 \cdot 6 = 336$
(iii)   10**P**4 $= 10 \cdot 9 \cdot 8 \cdot 7 = 5040$

So, given a set of n items, we can find the number of permutations obtained whether
(a) we include all the n items in each arrangement—n**P**n.

or

(b) we include only *r* of the items at a time—*n***P***r*.

Now how do these two numbers (*n***P***n* and *n***P***r*) compare in size?
For example, which is *larger*: 4**P**4 or 4**P**2?

4**P**4 is larger than 4**P**2. (You get more arrangements by permuting 4 items
together than by permuting only 2 of them at a time.)
$$4\mathbf{P}4 = 4! = 4\cdot3\cdot2\cdot1 = 24$$
$$4\mathbf{P}2 = \quad = 4\cdot3 \quad = 12$$

As an illustration, suppose our four items are A, B, C, and D. Consider each of
the 12 possible permutations of just 2 at a time:

$$
\begin{array}{cccc}
A\,B & B\,A & C\,A & D\,A \\
A\,C & B\,C & C\,B & D\,B \\
A\,D & B\,D & C\,D & D\,C
\end{array}
$$

Each of these permutations of 2 (e.g. A B) leaves 2 items (C and D) unused. To
make a permutation of 4 items, these 2 extra items could be added to A B in
two different ways (C D or D C), giving *twice* as many permutations. And the
same is true for each of the 12 permutations.

We've noticed that 4**P**2 = 12 and 4**P**4 = 24; now for a really *important*
point.

What is the value of $\dfrac{4\mathbf{P}4}{(4-2)!}$?

$$\frac{4\mathbf{P}4}{(4-2)!} = \frac{4!}{2!} = \frac{4\cdot3\cdot\cancel{2}\cdot\cancel{1}}{\cancel{2}\cdot\cancel{1}} = 4\cdot3 = 12 = 4\mathbf{P}2$$

So $4\mathbf{P}2 = \dfrac{4\mathbf{P}4}{(4-2)!}$

What makes 4**P**2 smaller than 4**P**4 is the lack of those extra permutations that
could be made with the items we have NOT taken. So, if we divide 4! by
$(4-2)!$ we cancel out the permutations that would arise from the two letters
we are not using. Thus:

$$4\mathbf{P}2 = \frac{4\mathbf{P}4}{(4-2)!}$$

Now let's see if this works with other examples. For instance, if it does work,
then

$$5\mathbf{P}2 = \frac{5\mathbf{P}5}{?}$$

By what quantity must we divide 5**P**5 in order to get 5**P**2?

$5\mathbf{P}2 = \dfrac{5\mathbf{P}5}{(5-2)!} = \dfrac{5!}{3!}$ is what we'd expect.

And $\dfrac{5!}{3!} = \dfrac{5\cdot4\cdot\cancel{3!}}{\cancel{3!}} = 20$

which is indeed the correct number of permutations of 5 things taken 2 at a time. Notice again that the lower part of the fraction is cancelling out the 3×2 ways that the three untaken items could combine with the existing permutations of 2 items.

Try another one:

$$10P4 = \frac{10!}{?}$$

What do we need as the denominator of this fraction?

We need the number of permutations that would have been obtained from the untaken items: $(10 - 4)! = 6!$

Thus: $10P4 = \dfrac{10!}{6!} = \dfrac{10 \cdot 9 \cdot 8 \cdot 7 \cdot \cancel{6} \cdot \cancel{5} \cdot \cancel{4} \cdot \cancel{3} \cdot \cancel{2} \cdot \cancel{1}}{\cancel{6} \cdot \cancel{5} \cdot \cancel{4} \cdot \cancel{3} \cdot \cancel{2} \cdot \cancel{1}} = 336.$

So this relationship between nPn and nPr works out correctly once again. In fact, it is true of all possible permutations. Let's try to write it out as a general formula.

We can say that the number of permutations of n different items taken r at a time is

$$nPr = \frac{n!}{(\quad)!}$$

How would you complete the formula?

$$nPr = \frac{n!}{(n - r)!}$$

Make a note of this panel.

> The number of PERMUTATIONS of n different items taken r at a time (without repetitions) is
>
> $$nPr = \frac{nPn}{(n - r)!} = \frac{n!}{(n - r)!}$$

(The $(n - r)!$ chops off the tail of the factorial and thus gets rid of unwanted permutations.)

Evaluate the following expressions, which you've seen before, setting out your work with the formula above:
- (i)   8P3
- (ii) 6P2
- (iii) 9P7   (Use your table of factorials for this one.)

(i) $8P3 = \dfrac{8!}{(8 - 3)!} = \dfrac{8!}{5!} = \dfrac{8 \cdot 7 \cdot 6 \cdot 5!}{5!} = 336$

(ii) $6P2 = \dfrac{6!}{(6 - 2)!} = \dfrac{6!}{4!} = \dfrac{6 \cdot 5 \cdot 4!}{4!} = 30$

(iii) $9\mathbf{P}7 = \dfrac{9!}{(9-7)!} = \dfrac{9!}{2!} = \begin{array}{c}\text{(Using your table}\\\text{of factorials)}\end{array} \dfrac{362\,880}{2} = 181\,440$

Now, without looking at your notes:
write out the formula for the permutations of *n* items taken *r* at a time:
$$n\mathbf{P}r = ?$$

$$n\mathbf{P}r = \dfrac{n!}{(n-r)!}$$

That is:
Divide factorial *n* by the factorial of the *difference* between *r* and *n*. This chops off the unwanted tail of *n*! thus getting rid of all the permutations that would have been available if all *n* items had been permuted at once.

Permutations of Things that are NOT All Different

So far we have only tried to permute things that are all different from one another. How do we go about it if, instead, we want to permute a set of objects *some* of which are *identical*? How will the number of permutations be affected?

For instance, here are two words, each with the same number of letters:

O L D
O D D

Write out all the permutations you can make from the letters of each world in turn.

<u>OLD</u>   ODL   LDO   LOD   DLO   DOL—6 permutations
<u>ODD</u>   DDO   DOD   —3 permutations

Because two of the letters in the second word are the same, you have lost the extra permutations you would have had if the letters had been different. If some of the objects in a set cannot be told apart, the number of possible permutations is *reduced*.

Here are two more words each with the same number of letters:

A D V E R T
A S S E S S

Which word would give you *fewer* permutations? (Don't try to count them.)

ASSESS would give fewer permutations than ADVERT. Swop around *any* two letters in ADVERT and you get a new permutation; swop around any two of the four Ss in ASSESS and the arrangement appears unchanged.

So, if a group contains any identical objects, it won't give rise to as many permutations as it would if all the objects could be told apart.

How, then do we calculate the number of permutations when not all the items are different? Let's look a bit more closely at this 6-letter word:

A S S E S S

How do we decide the number of possible arrangements? Clearly this would be easy if all the letters were different (as in ADVERT where D, V, R, and T take the place of all those Ss. How many permutations would be possible if all 6 letters *were* different?

There would be 6! permutations if all the 6 letters were different (as in ADVERT).

Now the six letters of the word ASSESS *can* be put in a number of different orders, e.g.

SASESS
SAESSS
ESSSSA
SSSSEA   etc., etc.,

But we don't, at the moment, know how many different orders are possible. Until we've solved the mystery, let's just use x to stand for the total number of possible permutations.

Will x be larger or smaller than 6!?

---

x (the number of permutations of 6 letters, 4 of them identical)
will be SMALLER than
6! (the number of permutations if all 6 letters are different).

Now exactly *how much* smaller? Let's look at just one of the x possible permutations:

| S | | S | | S | | S | | E | | A |

Suppose we alter the Ss so we can tell them apart—we'll attach numbers to them:

S1 S2 S3 S4 E A

Now we have four clearly *different* Ss together with an E and an A. Without moving the E and A from the right-hand end, we can make more permutations by swopping places among the Ss, e.g.:

S2 S4 S1 S3 E A

or

S4 S1 S3 S2 E A   etc., etc.

How many permutations are possible (including those above) if you swop the Ss around among the four places available for them?

There are 4! possible permutations.

With the E and the A in the position shown, the four other 'places' can be filled in $4 \cdot 3 \cdot 2 \cdot 1 = 4!$ ways by swopping the Ss which we've numbered so as to tell them apart.

The same thing would happen with *each and every one* of the x permutations of the letters in ASSESS (e.g. AESSSS or ASSSSE); if the Ss were made to look different, so we could tell them apart, we would get 4! permutations instead of the one.

So the word ASSESS gives x permutations, each with 4 letters identical. Each one of these permutations could be turned into 4! permutations by replacing the 4 identical letters with 4 different letters.

This would give us $\underline{\ ?\ } \times \underline{\ ?\ }$ permutations.

This would give us $x \times 4!$ permutations.

Now where have we got to?
—We started out with 6 letters, 4 of them the same.
—We decided the number of permutations should be called x.
—Then we tried making all 6 letters different.
—Making all the letters different has given us
$$x \times 4! = x(4!) \text{ permutations.}$$
Since the 6 letters are now all different, $x(4!)$ must be equal to the number of permutations obtained from *6 things that are all different*. (And you've calculated this before.)

So, $x(4!) = \underline{\ ?\ }$

So, $x(4!) = 6!$

Just be clear about what this equation says. It says 'if you multiply the number of permutations obtained from 6 items, 4 of which are alike (x) by the number of permutations you would get by permuting those like things among themselves (4!), you get the total number of permutations available from a set of 6 objects which are all different'.

So now you should be able to calculate x. How many ways can you arrange the letters of the word ASSESS (where 4 of the 6 letters are alike)?

$$\text{Since } x(4!) = 6!$$
$$x = \underline{\ ?\ }$$

Since $x\,(4!) = 6!$
$$x = \frac{6!}{4!} = \frac{6 \cdot 5 \cdot 4!}{4!} = 30$$

The letters of the word ASSESS can be permuted in 30 ways (as compared with the 6! permutations of the word ADVERT).

So far we've worked with just the one example of a set of items some of which are alike. Let's put our findings into more general terms:

From *n* items, *a* of which are alike, we get *x* permutations.
Change the *a* alike into a different items, and we get
$$x(a!) \text{ permutations.}$$
BUT this is the *same* as the number of permutations we'd get from *n* different items—*n*!
Therefore: $\quad n! = x(a!)$
$\qquad$ and $x = \underline{\ ?\ }$

Therefore: $\qquad\qquad\qquad n! = x(a!)$
$$\text{and} \quad x = \frac{n!}{a!}$$

To sum this up: the number of permutations of *n* items, *a* of which are alike, is naturally *smaller* than the number *n*! which you would get if all the items were different. In fact, it is smaller by the factor *a*! because this is the number of permutations you would get from *a* different items and which you are *losing* because they are all the same.

Thus $\qquad\qquad\qquad\qquad x = \dfrac{n!}{a!}$

So much for the explanation. Try an example:
How many permutations can you make using all the letters of the word PUPPY?

If *x* is the number of permutations obtainable from the 5 letters of PUPPY and *a* is the number of identical letters (the P's) then:
$$x = \frac{n!}{a!}$$
$$= \frac{5!}{3!} = \frac{5 \cdot 4 \cdot 3!}{3!} = 5 \cdot 4 = 20$$

So much for a group of items where some are alike. But what if the group contains more than one lot of identical items, e.g. how many permutations from the word MISSISSIPPI? You probably won't be surprised to learn that we can extend our formula to cope with this problem:

> If you have *n* things, containing *a* alike of one kind, *b* alike of a second kind, *c* alike of a third kind (and so on . . .); then the number of permutations of the *n* things taken all together is:
> $$\frac{n!}{a!\,b!\,c! \ldots}$$

Make a note of this panel. (If you wish to see how the formula can be reasoned out, look at page 106).
Now use the formula to answer the following question:

How many permutations can you make from the letters of the word ASSASSIN?

The word ASSASSIN has 8 letters, including 2 alike of one kind (As) and 4 alike of another kind (Ss). Therefore the number of permutations obtainable is:

$$\frac{8!}{4!2!} = \frac{8 \cdot 7 \cdot 6 \cdot 5 \cdot \cancel{4!}}{\cancel{4!}2!} = 840$$

Try these examples:

(i) In how many ways can I plant a line of 10 bulbs consisting of 5 tulips, 3 daffodils, and 2 narcissi?

(ii) I have 2 copies of a statistics textbook, 2 copies of an algebra textbook, 3 copies of a calculus textbook, and 1 copy of a textbook on astronomy. In how many ways can I allocate these among eight students so that each student has one textbook?

(iii) A boat has 6 signal flags:

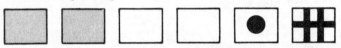

How many different signals can it make by arranging all 6 flags on a line?

(i) $\dfrac{10!}{5!3!2!} = \dfrac{10 \cdot 9 \cdot 8 \cdot 7 \cdot 6 \cdot 5!}{5!(3 \cdot 2 \cdot 1)2 \cdot 1} = 2520$

(ii) $\dfrac{8!}{2!2!3!} = \dfrac{8 \cdot 7 \cdot 6 \cdot 5 \cdot 4 \cdot 3!}{2 \cdot 1(2 \cdot 1)3!} = 1680$

(iii) $\dfrac{6!}{2!2!} = \dfrac{6 \cdot 5 \cdot 4 \cdot 3 \cdot 2 \cdot 1}{2 \cdot 1(2 \cdot 1)} = 180$

If you want to know how the formula $\dfrac{n!}{a!b!c! \ldots}$ can be reasoned out, look at the panel below. If not, go straight on to the next section when you are ready.

You want to know how this formula

$$\frac{n!}{a!b!c! \ldots}$$

can be reasoned out. Look at it like this:

We are trying to find the number of permutations obtainable from a group of n things taken all together, when there are a alike of one kind, b alike of a second kind, c alike of a third, and so on.

Suppose there are x such permutations. Consider any one of them. Change the a alike things into a different things. Then permute them among themselves without altering the position of the rest of the things. This gives rise to a permutations where there was only one before. This will happen with each of the x permutations—so we have $x(a!)$ permutations.

Combination			Permutations					
A	B	C	ABC	ACB	CAB	CBA	BAC	BCA
A	B	D	ABD	ADB	DAB	DBA	BDA	BAD
A	C	D	ACD	ADC	DAC	DCA	CAD	CDA
B	C	D	BCD	BDC	DBC	DCB	CBD	CDB

Notice that the four original combinations of 3 agents have each given rise to (*how many?*) permutations.

Each set of 3 agents can be arranged in 3.2.1 ways—*3!* permutations.

Now, just as we used 4P3 to stand for the permutations of 4 objects taken 3 at a time, so we can represent *combinations* of 4 objects taken 3 at a time by 4C3.

It is obvious that the 24 permutations of 4 agents taken 3 at a time, arise from re-arranging the combinations, i.e.

$$\text{Number of combinations} \times \text{ways of arranging each one} = \text{Number of permutations}$$

Or, in symbols:

$$4C3 \times 3! = 4P3$$

$$\text{Therefore } 4C3 = \frac{4P3}{?}$$

How would you complete this equation?

$$4C3 \times 3! = 4P3$$

$$\therefore 4C3 = \frac{4P3}{3!}$$

$$4C3 = \frac{4 \cdot 3 \cdot 2 \cdot 1}{3 \cdot 2 \cdot 1} = 4$$

Using this result we can very easily work out a formula for calculating the number of combinations available in any situation:

—The number of combinations you could make from *n* objects taking *r* of them at a time is *n*C*r*.

—Each of these *n*C*r* combinations would give *r*! permutations.

—And *n*C*r* multiplied by *r*! would give the total number of permutations available from *n* objects taken *r* at a time, i.e. *n*P*r*

—So, $nCr \times r! = nPr$

$$nCr = \frac{nPr}{r!}$$

Now, since we know $nPr = \dfrac{n!}{(n-r)!}$, what is the complete formula for *n*C*r*? *Write it down.*

$nCr = \dfrac{n!}{(n-r)!r!}$ Dividing *n*! by $(n-r)!$ cuts out the permutations that

Now change the *b* alike things into *b* different things. These can be permuted in *b*! ways and each can be linked with any of the $x(a!)$ permutations. So we now have $x(a!b!)$ permutations.

This can also be done with *c* alike things, and so on, with the result that if all the things are made different we have $x(a!b!c! \ldots)$ permutations.

But this is the number of permutations of *n* things when they are all different, that is *n*!

Therefore:
$$x(a!b!c! \ldots) = n!$$
$$\text{and} \quad x = \frac{n!}{a!b!c! \ldots}$$

From Permutations to Combinations

If we call a set of items a combination (rather than a permutation) it means we are *not* interested in the order in which those items are arranged. Thus, while *xyz*, and *yxz*, and *zxy* are different permutations, they are exactly the *same* combination.

For instance, at Counter-Espionage HQ we are choosing three agents from the four who are available to be sent across the East German border on a special mission. We have not yet decided whether to:
(i) send all three agents across the border together, or
(ii) send them separately at 8-hour intervals.

In which of these two cases are we considering combinations only; in which case permutations also?

In choosing 3 agents from 4 to be sent across the border *together*, order is not important—so we are considering combinations only. In fact, there are four such combinations.

Agents	A	B	C	(There are 4 combinations
or	A	B	D	because there are 4 ways of
or	A	C	D	dropping an agent from the
or	B	C	D	selection)

But if we are to send the selected agents across the border at 8-hour intervals, we have permutations to worry about. What order shall they cross in?

Clearly, four agents taken three at a time can be permuted in 4P3 ways. So how many ways can we send three agents chosen from four?

$$\textbf{4P3} = \frac{n!}{(n-r)!} = \frac{4!}{(4-3)!} = \frac{4 \cdot 3 \cdot 2 \cdot \cancel{1}}{\cancel{1}} = 24$$

There are 24 ways of choosing 3 agents from 4 and sending them across the border one after another. These are the possible orders in which they may cross:

would have been available from those objects we are not taking at all; and dividing by $r!$ cuts out the permutations that could be made from each combination we do take.

The Combinations Formula

> The number of COMBINATIONS from a set of n different items, taken r at a time, is
> $$n\mathbf{C}r = \frac{n\mathbf{P}r}{r!} = \frac{n!}{(n-r)!r!}$$

Make a note of this panel.

Now let's get some practice.

Two ambulance attendants are needed in an emergency and there are six men on duty. In how many ways can the two be chosen?

There are 15 ways of choosing 2 men from 6.
$$n\mathbf{C}r \frac{n!}{(n-r)!r!} = \frac{6!}{4!2!} = \frac{6 \cdot 5 \cdot \cancel{4!}}{\cancel{4!}2!} = 15$$

Another:

A party of eight people arrive at a river and a boatman agrees to take them across in two groups of four. In how many ways can the first boatload be chosen?

$$n\mathbf{C}r = \frac{n!}{(n-r)!r!}$$
$$8\mathbf{C}4 = \frac{8!}{4!4!} = \frac{8 \cdot 7 \cdot 6 \cdot 5 \cdot \cancel{4!}}{\cancel{4!}4 \cdot 3 \cdot 2 \cdot 1} = 70$$

Now try these:

(i) Calculate 14C12, 10C9, 100C98, 5000C5000.

(ii) A magazine called *Motor Choice* lists four 'best buys' from a range of twelve sports cars. In the next issue it lists three 'best buys' from among fifteen family cars. In which issue did the editors have the greater number of possible combinations to choose from?

CHECK your answers:

(i) $14\mathbf{C}12 = \dfrac{14!}{12!2!} = \dfrac{14 \cdot 13 \cdot 12!}{12!2!} = 91$

$\quad 10\mathbf{C}9 = \dfrac{10!}{9!1!} = \dfrac{10!9!}{9!1!} = 10$

$\quad 100\mathbf{C}98 = \dfrac{100!}{98!2!} = \dfrac{100 \cdot 99 \cdot 98!}{98!2!} = 4950$

$$5000C5000 = \frac{5000!}{5000!0!} = \frac{5000!}{5000!} = 1$$

(ii) Four choices from twelve gives 12C4 combinations.
Three choices from fifteen gives 15C3 combinations.

$$\text{Since } 12C4 = \frac{12!}{8!4!} = \frac{12 \cdot 11 \cdot 10 \cdot 9 \cdot 8!}{8!4 \cdot 3 \cdot 2 \cdot 1} = 495$$

$$\text{and } 15C3 = \frac{15!}{12!3!} = \frac{15 \cdot 14 \cdot 13 \cdot 12!}{12!3 \cdot 2 \cdot 1} = 455$$

So the magazine had more combinations to consider in the issue when they chose four sports cars from twelve.

Now try these.

(iii) In a local police station, a different combination of three policemen is selected daily for traffic duty. If there are ten policemen available, and the system operates every day except Sunday, how many weeks will go by before the same three policemen are on traffic duty again together?

(iv) I have eight friends at work. If I hold a party, how many ways can I invite
    (a) 5 of them?
    (b) 3 of them?

(iii) $10C3 = \dfrac{10!}{7!3!} = \dfrac{10 \cdot 9 \cdot 8 \cdot 7!}{7!3!} = 120$

So there are 120 different combinations (of three policemen chosen from ten), and 6 of these combinations are used each week. So it will be $\dfrac{120}{6} = 20$ weeks before the same three policemen are on traffic duty together again.

(iv) (a) $8C5 = \dfrac{8!}{5!3!} = \dfrac{8 \cdot 7 \cdot 6 \cdot 5!}{5!3!} = 56$

    (b) $8C3 = \dfrac{8!}{3!5!} = \dfrac{8 \cdot 7 \cdot 6 \cdot 5!}{3!5!} = 56$

Did that last example surprise you? All it means is this: each time we make our choice of items to include in a combination we are also 'choosing' a set of items to *reject* from that combination. Thus, each time we choose a combination of 5 items from 8 items it leaves a remainder—the rejected combination of 3 items. Similarly, for each combination of 3 items we choose, we reject a combination of 5 items.

    However many combinations of r objects we can make from n objects, there is an equal number of combinations each containing (*how many?*) objects.

... containing $n - r$ objects.

So it is quite easy to use the nCr formula to help you select *two* combinations at once. Our next example will show how.

Forming Groups

In how many ways can 7 different books be given to two students if one student is to have 4 books and the other is to have 3 books?

In selecting 4 books for one student you are leaving 3 for the other (or vice versa). So the number of ways of getting these combinations is

$$7C4 \times 7C3 = \frac{7!}{4!3!} = \frac{7 \cdot 6 \cdot 5 \cdot 4!}{4!3!} = 35$$

But what if we need *more than two* combinations? For instance, suppose we have 10 social workers and we wish them to work in groups of 2, 3 and 5. How many ways can we split them up into groups of the right size.?

Firstly, the group of 2 can be chosen in $10C2 = 45$ ways. Out of the remaining 8 workers we can select 3 in 8C3 ways, leaving the remaining 5 workers to form the third group.

So the first group can be chosen in 45 ways and for each of these ways there are $8C3 = 56$ ways of forming the other two groups.

So how many ways altogether of splitting 10 workers into groups of 2, 3 and 5?

There are $10C2 = 45$ ways of choosing 2 workers from 10; and there are $8C3 = 56$ ways of splitting up the remaining 8 workers into groups of 3 and 5. Since each of the 56 'split-ups' could be linked with any of the 45 selections that were possible originally, the total number of ways of forming all three groups is

$$10C2 \times 8C3 = 45 \times 56 = 2520$$

Try another:
In this year's school play the cast consists of 5 members of a royal family, 15 soldiers and courtiers, and 30 peasants and revolutionaries. How many ways can the cast of 50 pupils be divided among these three groups? (Choose the peasants first, then the soldiers etc., and leave your answer in *nCr* form.)

There are 50C30 ways of choosing groups of 30 from among 50 pupils, and for each of these there are 20C15 ways of splitting up the remaining 20 into groups of 15 and 5. Thus, there are 50C30 × 20C15 ways of making the three groups. (There are other equally correct answers such as 50C5 × 45C30 or 50C15 × 35C5 etc.)

Combining Combinations

Quite often the number of ways of dealing with a problem is found by combining the ways of forming two different combinations. For instance, suppose we have to form a committee of two boys and three girls from a group of 4 boys and 6 girls.

How many ways can we choose the boys? → 4C2 = 6
How many ways can we choose the girls? → 6C3 = 20
    Each combination of girls can be linked with each combination of boys, so how many different committees could we form?

___

6 × 20 = 120 committees.

Just check through these steps again:
1   Form a committee of 2 boys and 3 girls from a group of 4 boys and 6 girls.
2   Choose the boys in 4C2 = 6 ways.
3   Choose the girls in 6C3 = 20 ways.
4   This gives 6 × 20 = 120 ways of combining the two combinations.
(Notice how this ties in with the work we did on combining *probabilities* in Chapter 2—the choice of boys is *independent* of the choice of girls, and the overall result is found by *multiplying*.)
    Try another:
A basketball coach must select two guards and two forwards from among three guards and five forwards. How many different combinations of guards and forwards can he select?

___

$$3C2 \times 5C2 = \frac{3!}{2!1!} \times \frac{5!}{3!2!} = \frac{5!}{2!2!1!}$$
$$= \frac{5 \cdot 4 \cdot 3 \cdot 2 \cdot 1}{2 \cdot 2 \cdot 1} = 30 \text{ different combinations}$$

    Another:
Among his more valuable specimens, a stamp collector has 9 rare Polish stamps, 12 rare Hungarian stamps, and 6 rare Bulgarian stamps. As a 21st birthday present he wishes to give a third of each set of stamps to his nephew. How many ways can he make up this birthday packet? (Take the stamps in the order listed and leave your answer in *nCr* form.)

___

He can make up his birthday packet in 9C3 × 12C4 × 6C2 ways.

Again:
A soccer team of 11 players is to be chosen from 30 boys, of whom 4 can play only in goal, 12 can play only as forwards, and the remaining 14 in any of the other positions. If the team is to include five forwards and, of course, one goalkeeper, in how many ways can it be made up? (Leave your answer in *nCr* form.)

___

There are:
    4C1 ways of choosing the goalkeeper
    12C5 ways of choosing the forwards
and 14C5 ways of choosing the other 5 players.
That is, 4C1 × 12C5 × 14C5 combinations altogether.

One more:

A boy has a collection of 30 LP records—18 of rock groups, and 12 of folk singers. He is asked to bring a selection of 10 to a party. How many different selections could he make if

(i) he can bring any 10 he pleases?

(ii) the selection must include 8 rock and 2 folk?

(iii) the selection must include 3 particular folk LPs?

(Leave your answer in *n*C*r* form.)

The boy has 30 LPs (18 rock and 12 folk) and must choose 10:

(i) Any 10: 30C10 possible combinations

(ii) Including 8 rock and 2 folk: 18C8 × 12C2 combinations

(iii) Including 3 particular folk records: 27C7 (Since three of his records are already chosen for him, he has 7 to choose from the remaining 27 LPs.)

We have already seen that there are 18C8 × 12C2 ways in which the boy could select 8 rock LPs and 2 folk LPs. Now suppose he were asked to include *at least 8* rock LPs in his selection? In addition to the 18C8 × 12C2 ways of selecting exactly 8 rock LPs, we would have to consider the number of ways of selecting 9 rock and 1 folk, and 10 rock and no folk.

So, how many ways are there of selecting at least 8 rock LPs among the 10 LPs he chooses from 30? (Leave your answer in *n*C*r* form.)

There are (18C8 × 12C2) + (18C9 × 12C1) + 18C10 ways. That is, we add together the ways of getting 8 rock and 2 folk

9 rock and 1 folk

10 rock and 0 folk

(If you recall the language we used in Chapter 2, these possible selections are *mutually exclusive* alternatives.)

Try this question:

A football team of 11 men is to be chosen from 15 men. Four of the 15 can receive and two others can pass. If the team must include only one quarterback and at least three receivers, in how many ways can it be chosen?

(Notice that once you have considered the number of ways of forming a 3-receiver team you still have to consider the number of ways of forming a team with more than three receivers. Work your answer out to an exact number.)

We had to select 1 quarterback, 3 (or 4) receivers, and 7 (or 6) other players from 2 quarterbacks, 4 receivers, and 9 other players.

We can select a team with 3 receivers in

2C1 × 4C3 × 9C7 ways.

We can select a team with 4 receivers in

2C1 × 4C4 × 9C6 ways.

Thus the number of ways of selecting a team with either 3 or 4 receivers is 2C1 × 4C3 × 9C7) + (2C1 × 4C4 × 9C6)

$$= \left(2 \times 4 \times \frac{9 \cdot 8 \cdot 7!}{7!2!}\right) + \left(2 \times 1 \times \frac{9 \cdot 8 \cdot 7 \cdot 6!}{6!3!}\right)$$

$$= (2 \times 4 \times 36) + (2 \times 1 \times 84) = 456 \text{ different teams could be formed.}$$

Review Questions

Try these examples as a review of this section.

1   How many ways can a traffic policeman with 3 parking tickets left in his book give them to three out of the seven cars he finds that have over-parked?

2   It is desired to allocate the seven parts in a school play to the members of a class consisting of 12 boys and 14 girls. In how many ways may this be done if three of the parts are to be allotted to the boys and four to the girls? Leave your answer in *nCr form*.

3   A bag contains 2 white and 3 red cubes, all of different sizes. In how many ways can 3 cubes be selected from the bag if (a) at least one cube is white, (b) at least one cube is red?

Answers to Review Questions

1 $7C3 = \dfrac{7!}{3!4!} = \dfrac{7 \cdot 6 \cdot 5 \cdot 4!}{3!4!} = 35$

2   12C3 × 14C4

3   (a) (2C1 × 3C2) + (2C2 × 3C1) = (2 × 3) + (1 × 3) = 9
   (b) (3C1 × 2C2) + (3C2 × 2C1) + (3C3) = (3 × 1) + (3 × 2) + 1) = 10

NOTE: Some textbooks use a symbol of the form $\dbinom{n}{r}$ or $^{n}C_r$ to stand for a combination, rather than *nCr*, which we have been using. Thus 12C3 could be written as $\dbinom{12}{3}$ or $^{12}C_3$. The meaning is exactly the same.

4

Probability by Combinations

Probability Again

It is now time we got back to the subject of this book which is probability. How do you use what you've learned about combinations to calculate probabilities? First let me remind you of the formula for theoretical probability:

$$\text{Theoretical probability of result we are looking for} = \frac{\text{No. of outcomes giving 'looked-for' result}}{\text{Total no. of equally likely outcomes}}$$

Thus, if we know there are 52 cards that could equally well be drawn from a pack (4 of them being aces), there are 52 equally likely outcomes; and if drawing an ace is the result we are looking for, 4 of the 52 outcomes will give it—therefore probability $= \frac{4}{52}$.

Suppose we have a box of 20 pills, 5 of them containing a new drug and the rest being made of sugar. If a patient takes a pill at random, what is the probability he chooses the new drug?

$$\text{Probability} = \frac{\text{No. of outcomes giving 'looked-for' result}}{\text{Total no. of equally likely outcomes}} = \frac{5}{20} = \frac{1}{4}$$

Here 5 out of the 20 equally likely outcomes give the result we are looking for—choosing the new drug. If we had been looking for the probability of choosing the sugar pills, 15 of the outcomes would have been of interest.

That was to remind you of how to use the formula to calculate the probability of a single result.

Now, to calculate the probability of two separate results happening together, you can use the *multiplication rule*. Thus, if our patient takes two pills instead of one, what is the probability that *both* are the new drug?

$$\text{Probability that both pills are the new drug} = \frac{5}{20} \times \frac{4}{19} = \frac{1}{19}$$

The Combinations Method

Now let's solve this same problem by combinations.
We have 20 pills in a box and 5 of them contain the new drug.

How many ways can we choose 2 pills from these 20? That is, how many combinations of 2 could we make?

We can choose 2 pills from 20 in
$$20C2 = \frac{20!}{18!2!} = \frac{20 \cdot 29 \cdot 18!}{18!2!} = 190 \text{ ways}$$
So there are 190 ways of selecting a pair of pills from our box of 20. But how many of these ways will give the result we are looking for—two pills each containing the new drug?

Well, there are 5 pills containing the new drug. So, in how many ways can we choose 2 of them? (How many combinations of 2 could come from these 5?)

We can choose 2 drug pills from 5 drug pills in
$$5C2 = \frac{5!}{3!2!} = \frac{5 \cdot 4 \cdot 3!}{3!2!} = 10 \text{ ways}$$
So there are 190 ways of choosing 2 pills from 20.
And 10 of these combinations of 2 will consist of 2 of the 5 drug pills.
So we can now use the formula:

$$\frac{\text{No. of ways (outcomes) giving looked-for result}}{\text{Total no. of equally likely ways (outcomes)}} = \frac{5C2}{20C2}$$

$$= \frac{10}{190}$$

$$= \frac{1}{19} \quad \text{AS BEFORE}$$

Use the combinations method to find the probability of choosing 2 *sugar* pills from the 20.

We have 20 pills, 15 of them being made of sugar.
We can choose 2 from the 15 in 15C2 ways = 105
We can choose 2 from the 20 in 20C2 ways = 190

Probability of
choosing 2 sugar pills $= \dfrac{\text{No. of ways of selecting 2 of the 15 sugar pills}}{\text{No. of ways of selecting any 2 of the 20 pills}}$

$$= \frac{15C2}{20C2} = \frac{105}{190} = \frac{21}{38}$$

Apply the same combinations method to this probability problem:
If three letters are chosen at random from the word ASSESS, what is the probability that all three letters are Ss?

ASSESS has 6 letters, 4 of them being Ss:
We can choose 3 from the 4 in 4C3 = 4 ways.
We can choose 3 from the 6 in 6C3 = 20 ways.

$$\frac{\text{Probability of}}{\text{choosing 3 Ss}} = \frac{\text{No. of ways of choosing 3 of the 4 Ss}}{\text{No. of ways of choosing any 3 of 6 letters}}$$

$$= \frac{4C3}{6C3} = \frac{4}{20} = \frac{1}{5}$$

You can check this by the multiplication rule: $\frac{4}{6} \times \frac{3}{5} \times \frac{2}{4} = \frac{1}{5}$

In using this combinations approach to probability we are comparing the number of ways of getting the combination we are looking for (size *and* type) with the number of ways of getting a combination of the right size only.

Try another:

There are six men, two women, and two children travelling on a bus. If four people get out at the next stop, what is the probability they are all men?

There are 10 people on the bus, 6 of them being men. So:

$$\frac{\text{Probability that next}}{\substack{\text{4 people to get off}\\ \text{the bus are all men}}} = \frac{\text{No. of ways of choosing 4 men from 6}}{\text{No. of ways of choosing 4 passengers from 10}}$$

$$= \frac{6C4}{10C4} = \frac{15}{210} = \frac{1}{14}$$

Now let's suppose that Mrs XYZ is one of the 10 passengers on the bus. What is the probability that she will be one of the four passengers to alight at the next stop? Well, we know that there are 210 possible combinations of 4 that could get off the bus. How many of these will contain Mrs XYZ?

What we have to do before we can calculate this probability is decide in how many ways it is possible to 'choose' 4 people from 10 in such a way that a particular person (Mrs XYZ) is in the group. Can you see how this is done?

Since, in effect, 1 of your 4 passengers has already been chosen for you, what you have to do is choose _?_ passengers from the remaining _?_ passengers.

... choose *3* passengers from the remaining *9* passengers.

So, given that one named passenger is to be included in the 4,
we can choose the other 3 passengers in 9C3 = 84 ways, and
we can choose 4 passengers from 10 in 10C4 = 210 ways.

$$\frac{\substack{\text{Probability}\\ \text{that Mrs}\\ \text{XYZ is}\\ \text{one of 4}}}{} = \frac{\text{No. of ways of choosing 3 other passengers from the other 9}}{\text{No. of ways of choosing 4 passengers from 10}}$$

$$= \frac{9C3}{10C4} = \frac{84}{210} = \frac{2}{5}$$

(If you want to check this, calculate the probability that Mrs XYZ would NOT be one of the first four to get off, and then use $p = 1 - q$.)

Use our 'combination method' on this one*:
The eleven players in a soccer team are drawing names from a hat to decide which of them will represent their team at 5-a-side soccer. What is the probability that the regular captain will be a member of the team selected?

What we are asking is how many ways can we select 4 other players besides the captain from a group of 10 players, and how many ways could we select 5 from 11 without any restriction?

$$\text{Probability} = \frac{\text{No. of ways of selecting 4 others from 10}}{\text{No. of ways of selecting 5 players from 11}}$$

$$= \frac{10C4}{11C5}$$

$$= \frac{\dfrac{10!}{6!4!}}{\dfrac{11!}{6!5!}} \longleftarrow$$ (Notice how we can simplify the working by inverting the denominator of the probability fraction, and then cross-multiplying.)

$$= \frac{10!}{6!4!} \times \frac{6!5!}{11!}$$

$$= \frac{\cancel{10!}}{\cancel{6!4!}} \times \frac{\cancel{6!}\, 5 \cdot \cancel{4!}}{11 \cdot \cancel{10!}} = \frac{5}{11}$$

Now see if you can extend this idea a shade further:

What is the probability that both the regular captain *and* the regular goalkeeper will be among the five players selected from eleven?

$$p = \frac{9C3}{11C5}$$

If the captain and goalkeeper are to be included in the 5 it means we are left to select 3 players from the remaining $11 - 2 = 9$ players. Therefore:

$$\text{Probability} = \frac{\text{No. of ways of selecting 3 others from 9}}{\text{No. of ways of selecting 5 from 11}}$$

$$= \frac{9C3}{11C5}$$

$$= \frac{\dfrac{9!}{6!3!}}{\dfrac{11!}{6!5!}} = \frac{9!}{6!3!} \times \frac{6!5!}{11!} = \frac{2}{11}$$

*Yes, I know you can see at a glance that the probability is $\dfrac{5}{11}$, but we need to test the 'combination method' on some 'simple' problems before we go on to those it is really essential for.

A boy is fishing on a pier and in his tackle box he has 3 cod-hooks, 5 flounder-hooks, and 2 eel-hooks. His box gets over-turned and half his hooks drop into the water. What is the probability he has lost *all* his cod-hooks?

The boy starts with 10 hooks and loses 5.
The 5 can be made up in 10C5 ways.
If 3 of them are cod-hooks (that is, *all* his cod-hooks) the other 2 can be made up from the remaining 7 hooks in 7C2 ways. Therefore:

$$\frac{\text{Probability he has lost all}}{\text{3 cod-hooks among the 5}} = \frac{\text{No. of ways of losing 2 other hooks from 7}}{\text{No. of ways of losing 5 hooks from 10}}$$

$$= \frac{7C2}{10C5}$$

$$= \frac{7!}{2!5!} \times \frac{5!5!}{10!} = \frac{1}{12}$$

With these two examples, don't work out your answer fully—leave them in *n*C*r* form.

(i) A patrol of 18 soldiers is to be chosen from 90. What is the probability that two particular soldiers will be chosen?

(ii) In a survey designed to find out how far local housewives travel to do their shopping, a sample of 1000 is chosen at random from a population of 20 000 women. What is the probability that both Mrs Rose and her friend Mrs Ruby will be chosen for the sample?

(i) $p = \dfrac{88C16}{90C18}$

(ii) $p = \dfrac{19\,998C998}{20\,000C1000}$

As you will have realised, most of the probabilities we've worked out with combinations so far could just as well have been found by using the 'equally likely outcomes' formula. Let's move on now to problems that could not so easily be solved by the old formula.

Suppose the treads on my car tires are worn below the legal limit and I must buy 4 new tires. A rather shady dealer offers me (cheap) a choice from a pile of 9 new tires, 3 of which (unknown to me) contain a severe fault. What is the probability I will choose 2 good tires and 2 faulty ones?

Let's take this in stages:
First of all, how many selections of 4 can I make from the 9 tires?

I can make 9C4 selections of 4 tires from the 9 available.

So I have 9 tires to choose from and 3 are faulty (though the faults are not obvious). What is the probability I'll choose 2 good, and 2 faulty tires? Altogether, there are 9C4 possible combinations of 4 tires that I could choose.

So, next, in how many ways could I choose 2 *good* tires?

I could choose 2 good tires (from the 6 good tires in the pile) in 6C2 ways.

Next, from the 3 faulty tires in the pile, how many ways could I choose 2 faulty ones?

There are 3C2 ways of choosing 2 faulty tires from 3.

So I know that there are
        6C2 = 15 ways of choosing 2 good tires
and   3C2 = 3 ways of choosing 2 faulty tires
    Each of the 15 combinations of 2 good tires could be linked with any of the 3 combinations of 2 faulty tires, giving   ?   possible combinations of 2 good and 2 faulty.

... giving *45* (i.e. 15 × 3) possible combinations of 2 good and 2 faulty.

So, to sum up:
    There are 6C2 × 3C2 ways of selecting 2 good and 2 faulty tires from 9 tires, 6 of which are good and 3 faulty. And altogether there are 9C4 ways of selecting 4 tires. So:

$$\text{Probability of choosing 2 good and 2 faulty tires} = \frac{6C2 \times 3C2}{9C4} = \frac{15 \times 3}{126} = \frac{45}{126} = \frac{5}{14}$$

Try another example:
Three policemen are needed for special duty but eight patrolmen and four sergeants have volunteered. To choose fairly, all twelve names are put in a helmet and three are drawn out. What is the probability we've chosen two patrolmen and one sergeant?

There are 8C2 = 28 ways of selecting 2 patrolmen from 8,
        and 4C1 = 4 ways of selecting 1 sergeant from 4.
Any of the 28 possible combinations of 2 patrolmen could be linked with any of the 4 possible sergeants, giving
            8C2 × 4C1 = 28 × 4 = 112
possible combinations of 2 patrolmen and 1 sergeant. Altogether, there are 12C3 = 220 ways of choosing 3 from 12, so

$$\text{Probability of choosing 2 patrolmen and 1 sergeant} = \frac{8C2 \times 4C1}{12C3} = \frac{28 \times 4}{220} = \frac{112}{220} = \frac{28}{55}$$

A driver has, in his 'spares'-box, 3 Champion spark plugs, 5 Autolite spark plugs, and 7 A.C. spark plugs. If he takes out 6 plugs at random, what is the probability he takes 2 of each kind? (Leave your answer in *n*Cr form.)

Our driver can choose
    the 2 Champion plugs from 3 in 3C2 ways
    the 2 Autolite plugs from 5 in 5C2 ways, and
    the 2 A.C. plugs from 7 in 7C2 ways.
And he can choose *any* 6 plugs from 15 in 15C6 ways. Therefore:

$$\frac{\text{Probability he chooses}}{\text{2 of each}} = \frac{3C2 \times 5C2 \times 7C2}{15C6}$$

Six Russian cosmonauts and five U. S. astronauts are in training for a joint excursion to Mars. The 5-man crew is to be chosen by picking names at random. What is the probability that the crew will consist of three Russians and two Americans? (Work it out in full.)

We have 6 Russians and 5 Americans from whom a crew of 5 must be chosen. The number of ways of choosing
    3 Russians from 6 is 6C3 = 20
    2 Americans from 5 is 5C2 = 10.
Any of the 20 combinations of Russians could be teamed with any of the 10 combinations of Americans, giving
    6C3 × 5C2 = 20 × 10 = 200 possible crews consisting of 3 Russians and 2 Americans.
Since the total number of ways of selecting any 5 men from 11 is 11C5 = 462:

$$\frac{\text{Probability of choosing 3}}{\text{Russians and 2 Americans}} = \frac{6C3 \times 5C2}{11C5} = \frac{200}{462} = \frac{100}{231}$$

Now let's take just one more step with this particular problem and link it to what you were doing a few pages back. Let's calculate the probability that certain *particular* men from the eleven will be chosen for the crew. For instance, what is the probability that Sergei (U.S.S.R.) and Hank (U.S.) will be in a crew of 3 Russians and 2 Americans?

First, would you expect this probability to be greater *or* less than $\frac{100}{231}$?

The probability of two particular men being in the crew is LESS than $\frac{100}{231}$.

That is, the three Russians and two Americans could be chosen for the crew in many ways, but not all such crews would contain Sergei and Hank.

So let's work out just what the probability is. What is the probability of choosing three Russians including Sergei, and two Americans including Hank, from the six Russians and five Americans available?

Clearly, if Sergei is to be one of the Russians, we are left to choose 2 Russians from the other 5. If Hank is to be one of the Americans we are left to choose 1 American from the other 4.

This gives us ___?___ combinations of 3 Russians, including Sergei, and 2 Americans, including Hank.

=====

... $5C2 \times 4C1 = 40$ combinations.

And since there are still $11C5 = 462$ ways we can choose a 5-man crew from 11 without restrictions.

$$\text{\textit{Probability of choosing Sergei and two other Russians and Hank and one other American}} = \frac{5C2 \times 4C1}{11C5} = \frac{40}{462} = \frac{20}{231}$$

Try another:
At the end of the last section I asked you how many ways the seven parts in a school play could be allocated between twelve boys and fourteen girls if three of the parts are for boys and four for girls. Now:

What is the probability that a particular girl and a particular boy will be in the cast? (Leave your answer in nCr form.)

=====

3 boys and 4 girls to be chosen from 12 boys and 14 girls.
Assuming 1 boy and 1 girl are already picked out, we can choose
   the other 2 boys from the remaining 11 in $11C2$ ways
   the other 3 girls from the remaining 13 in $13C3$ ways.
So there will be $11C2 \times 13C3$ combinations containing a particular girl and a particular boy among its 3 boys and 4 girls. But there are $26C7$ ways of choosing 7 people from the 26 available, so:

$$\text{\textit{Probability of choosing particular boy and 2 others and particular girl and 3 others}} = \frac{11C2 \times 13C3}{26C7}$$

A few more problems to finish off this section:
   (i) A box in a jeweller's office contains 20 diamonds, of which 8 are fakes. If a thief only has time to grab 5 of them before he is disturbed, what is the probability he has got nothing but fake diamonds? (Answer in nCr form.)
   (ii) A production line produces 20% defective items. What is the probability that in a random sample of 4 from a batch of 30, 3 will be defective? (nCr form only.)

=====

(i) $\dfrac{8C5}{20C5}$ (ii) $\dfrac{6C3 \times 24C1}{30C4}$

And now try one that perhaps needs a little more thought.
   (iii) In a local car race there are 25 entrants including 6 Corvettes. Assuming everyone has an equal chance of finishing in the first four

places, what is the probability the Corvettes will occupy at least 2 of the first 4 places? (Answer in *nCr* form only.)

(iii) $\dfrac{6C2 \times 19C2}{25C4} + \dfrac{6C3 \times 19C1}{25C4} + \dfrac{6C4}{25C4}$

(In this example you had to add together a set of mutually exclusive possibilities.)

You now have the tools and formulae to solve a great many probability problems by means of combinations. However, not all problems are obvious, and you sometimes have to do a fair bit of thinking round a situation before you can decide on the best method to use.

Review

Most of the problems we've looked at in this section can be summed up as follows:

Problem
What is the probability of selecting a sample containing so many items of type A and so many of type B (e.g. men and women, Russians and Americans, Corvettes and others, etc.) from a population containing a known number of each type?

Combinations Method
① In how many ways can we select a sample of the required total *size* from the population?
② In how many ways can we select the required number of type A items?
③ In how many ways can we select the required number of type B items?
(And we could continue to types C, D, E etc.).
④ Multiply ② and ③ to give total number of ways we can select a sample of required proportions.
⑤ Divide by ① to find the probability of getting that sample.

$$\frac{② \times ③}{①} = p$$

Make a note of this panel.

See how well you can cope with this problem I posed right at the beginning of Chapter 3.

The diagram overleaf shows a factory window. Every lunch-time the factory's workers play softball outside and they have broken 3 of its 25 panes. What is the probability that the three broken panes lie on a diagonal? (There are fourteen diagonals altogether.)

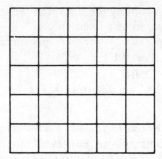

Give this problem some thought, then read on to see if your working agrees with mine.

Let's look at the problem step by step.

What is the probability that three broken panes lie on a diagonal? To answer this we need two pieces of data:

① How many different combinations of 3 broken panes can there be among 25?

② How many of these combinations of 3 will lie on a diagonal?

Divide ② by ① and you'll find the required probability.

① should give you no trouble, but how do we find ②?

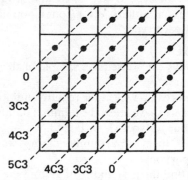

You'll see I have drawn in seven diagonals on the diagram and shown how many ways you could break 3 of the panes from each.

All you have to do now is

(a) Mark the other set of diagonals in the same way.

(b) Combine all the possible combinations of 3 to get the numerator of your probability fraction (② above).

(c) Divide ② by ①.

OK, so far? If you haven't done so already, carry out steps (a)–(c) above.

Altogether there are 40 combinations of 3 that lie on a diagonal.

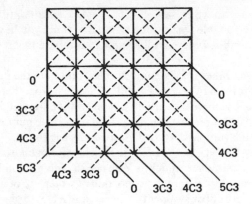

$$2(3C3 + 4C3 + 5C3 + 4C3 + 3C3) = 2(1 + 4 + 10 + 4 + 1)$$
$$= 2(20)$$
$$= 40$$

And there are $25C3 = 2300$ ways of breaking 3 out of 25 panes.
Therefore:

$$\text{\textit{Probability that 3 broken panes lie on a diagonal}} = \frac{\text{No. of ways of breaking 3 on a diagonal}}{\text{No. of ways of breaking 3 out of 25}}$$
$$= \frac{2(3C3 + 4C3 + 5C3 + 4C3 + 3C3)}{25C3}$$
$$= \frac{40}{2300} = \frac{2}{115}$$

Experiments in Probability and Sampling

If you can find time and equipment to do some of the following experiments it will help you considerably with your understanding of the remaining sections of this book.

1 Get a bag containing at least 100 marbles of which 25% are of a special colour distinct from the rest. Draw at least 128 samples of 4 at random (get a little help from your friends). Construct a table showing the number of 'special-coloured' marbles in each sample.

Replace the marbles in the bag between one draw and the next. (Later you'll be able to compare your actual results with the theoretical results.)

2 From the same bag as in Experiment 1, draw samples of 4, 8, 16, 20, 40, 60, 100 (replacing the marbles after each sample) and record the number of 'special-coloured' marbles in each sample. Use the number of specials in each sample as an estimate of the proportion in the population as a whole. How does the size of the sample affect the accuracy of this estimate? Does the average of 25 samples of 4 give you as good an estimate as one sample of 100?

3 Draw a card at random from a well-shuffled pack, return the card, and repeat three more times (or take one card from each of 4 packs). Note the number of spades in your sample of $n = 4$. Repeat until you have 100 samples. Draw up a table to show the number of times you have 0, 1, 2, 3, or 4 spades in your sample. Later you'll be able to compare theoretical and actual frequencies.

4 Toss a coin six times (or six coins at once) and note the number of heads you get in each sample of $n = 6$. Do this 128 times altogether and record the result each time. Later we'll compare actual and theoretical figures.

5 Throw five dice (or one die five times) and record the number of 6s. Repeat until you have 108 samples. Record your results each time.

6 Make a SAMPLING BOTTLE with which you can take large numbers of samples of various sizes very quickly. You need a jar big enough to hold 200 or more balls (or marbles), a known proportion of which are of a special colour. You need a cork or rubber stopper for the jar, and a closed glass tube just wide enough to let the balls through it and long enough to take up to, say, 20 balls.

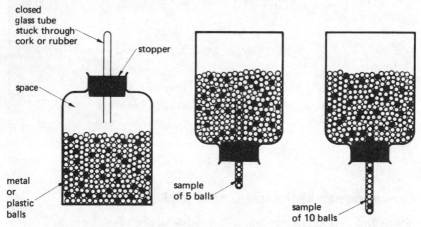

closed glass tube stuck through cork or rubber

stopper

space

metal or plastic balls

sample of 5 balls

sample of 10 balls

To take a sample from the population in the bottle, all you need do is tip the bottle up. To take samples of different sizes, move the glass tube in or out of the jar. Always shake the bottle well between samples.

Take a break at this point (and do some of the experiments if possible) before you continue.

Combinations and the Binomial Distribution

Binomial Probabilities

Combinations are particularly useful for calculating probabilities in a *binomial* situation. What does 'binomial' mean? Well, consider these situations.

A marksman hits the target or he misses.

A new-born baby will be a boy or a girl.

A space-rocket is launched successfully or it is not.

A manufactured article is sound or it is faulty.
A student either passes his examination or he fails.
A new drug cures a patient or it does not.
A customer buys Brand X soap-powder or some other.
An elector votes for the party in power or he does not.

What do these situations all have in common?

Each situation has *two* possible outcomes. Since we could name two possible results, such situations are called binomial (bi-nomial = 2 names).

As you might gather from the range of examples I gave, binomial situations are very common in everyday life. Whenever a situation or trial or experiment can have two possible results—whether or not they are equally likely—then it is binomial.

Suppose we have a large number of marbles (10% red and the rest white) in a barrel. If you put in your hand and draw out ONE marble at random:
(i) what are the two binomial possibilities?
(ii) are they both equally likely?

(i) The binomial possibilities are 'drawing a red' and 'drawing a white'.
(ii) The two possibilities are *not* equally likely, since only 10% of the marbles are red.

So, in taking a sample of 1 from the barrel of marbles we can choose red or white. And since 10% of the marbles are red:

$$p(\text{Red}) = \frac{1}{10}.$$

What happens if we take a sample of more than one marble, and don't replace those we take out from the barrel? Well, since there is a very large number of marbles, the probability of choosing a red marble stays virtually the same each time, and

sample of two: $p(\text{Red then Red}) = \frac{1}{10} \times \frac{1}{10} = \frac{1}{100}$

sample of three: $p(\text{Red then Red then Red}) = \frac{1}{10} \times \frac{1}{10} \times \frac{1}{10} = \frac{1}{1000}$

In general,

sample of n: $p(n \text{ Reds}) = \left(\frac{1}{10}\right)^n$

What is the probability of choosing n white marbles in a sample of n marbles from this barrel (where 90% of the marbles in it are white)?

$$p(n \text{ Whites}) = \left(\frac{9}{10}\right)^n$$

Each member of the sample of n has a $\frac{9}{10}$ chance of being white.

All this is perfectly easy and nothing new to you, also, you well know that the marbles chosen would rarely be all the same colour—either all red or all white. Most samples would contain *some of each* colour.

The real question is: what is the probability, in a sample of a certain size, of getting 0 red marbles, 1 red marble, 2 red marbles, 3 red, 4 red, 5, 6, and so on? This is a very practical question in manufacturing industry where batches of products are sampled and inspected and the proportion of faulty articles in the sample used to predict the proportion faulty in the output as a whole (the 'population').

Suppose a machine is manufacturing zippers with a defective rate of 25%—one zipper in four has some kind of fault. If we take a sample of 6 zippers from the large number produced, we could have in our sample 6, 5, 4, 3, 2, 1, or 0 faulty zippers. To take just two of these seven possibilities, which would you think *more* likely in your sample of 6:

　(a)　2 faulty?　　　　　or (b)　4 faulty?

(a) 2 faulty. There are fewer faulty than sound zippers in the output as a whole, so you are more likely to get a sample with a minority of faults than with a majority.

However, even though only a minority of the population are defective, *some* of your samples may suggest that a majority are. Often in industry the goods are inspected to make sure that the defective-rate is not changing: so what we want to know is:'if the proportion of defectives in the output as a whole is remaining constant, what is the probability of getting samples that suggest a higher or lower proportion?'

And if, for instance, we keep getting samples with a *higher* proportion of defective articles—*more often* than probability would lead us to expect—what might we conclude?

Surely we might well conclude that the proportion of defectives in the output as a whole is *higher* than we thought it was.

So, this is a real, practical question that arises in a good many scientific and industrial situations: 'if a single item can be of type A or type B, what is the probability that a sample of n such items will contain 0, 1, 2, 3, etc. of type A (or of type B)?' Let's start looking for answers.

As a simple example, let's say we have a huge box of marbles, 25% of them red and the rest white; and we have so many of these marbles that removing some of them will make no practical difference to the probabilities of choosing white or red.

If we take a sample of 1 marble:

　　　　Probability we get a White: $q = \dfrac{3}{4}$

　　　　Probability we get a Red:　$p = \dfrac{1}{4}$

Since Red and White are the only two possibilities $q + p \underline{\ ?\ }$.

$q + p = \underline{1}$

Now, if we take a sample of 2 marbles, there are 3 possibilities:

① Probability of choosing 0 Red $= q \times q = q^2 = \dfrac{3}{4} \times \dfrac{3}{4} = \dfrac{9}{16}$

② Probability of choosing 2 Red $= p \times p = p^2 = \dfrac{1}{4} \times \dfrac{1}{4} = \dfrac{1}{16}$

③ The only other possible result we could get in a sample of 2 is 1 Red and 1 White.

Since this is the only other possible result, the probability of choosing 1 Red (and 1 White) is $1 - (?) = \underline{?}$.

$1 - \dfrac{(1 + 9)}{16} = \dfrac{6}{16}$

So, when $p = \dfrac{1}{4}, q = \dfrac{3}{4}$, and we take a sample of $n = 2$, we have calculated the probabilities to be as follows:

$$0 \text{ Red} = \dfrac{9}{16} \qquad 1 \text{ Red} = \dfrac{6}{16} \qquad 2 \text{ Red} = \dfrac{1}{16}$$

We found the probability of the 'mixture' (1 Red, 1 White) by subtracting the probabilities of 'all or none' from 1. This would not be possible if we were dealing with samples of more than 2, so let me remind you of another method you've used before for finding such probabilities, e.g. $p(1 \text{ good egg, 2 bad})$.

On drawing a sample of 1, there are 2 possible outcomes—you draw a Red or you draw a White.

But if you take a sample of 2, there are more possible (joint) outcomes— you can draw a Red followed by Red, or (*what other possibilities?*).

Draw a probability tree and *list* the possible joint outcomes.

With a sample of 2, there are 4 possible joint outcomes:

1st	2nd		Joint outcome
R	R		RR
	W		RW
W	R		WR
	W		WW

So, with a sample of 2, the marbles you choose can be

Red followed by Red
Red followed by White
White followed by Red
White followed by White

(i) How many *permutations* do we have here?

(ii) How many *combinations*?

(i) There are 4 permutations: R → R, R → W, W → R, W → W.

(ii) There are 3 combinations: R and R, R and W, W and W.

The combination of 1 Red and 1 White can arise in two different ways (two different permutations). The probabilities of these two permutations are:

Probability of W → R $= q \times p = \left(\dfrac{3}{4} \times \dfrac{1}{4} = \dfrac{3}{16} \right) = qp$

Probability of R → W $= p \times q = \left(\dfrac{1}{4} \times \dfrac{3}{4} = \dfrac{3}{16} \right) = qp$

But BOTH these permutations give the SAME RESULT as far as we are concerned. We are not interested in the order in which the Red and the White are chosen—just so long as we have the right number of each, the right combination.

So the probability of 1 Red and 1 White (obtained in either order) can be simply written as

$$qp + qp = \underline{?}\,qp = \frac{?}{16}$$

$qp + qp = 2qp = \dfrac{6}{16}$ (And $\dfrac{6}{16}$ is, of course, the same result we got by subtraction a couple of pages ago.)

Now let's try to sum up the results we've got so far:

Sample of 1

Type of result	Probability
1 White	q
1 Red	p

$q + p = 1$

Sample of 2

Type of result	Probability
0 Red	q^2
1 Red (and 1 White)	$2qp$
2 Red	p^2

Look at the right-hand column of the 'Sample of 2' table.

The probability of getting either 0 or 1 or 2 Red marbles in a sample of 2 is
$$\underline{?} + \underline{?} + \underline{?} = \underline{?}$$

$$\ldots q^2 + 2qp + p^2 = 1$$

Now, if you've done a bit of algebra, you may well recognise that $q^2 + 2qp + p^2$ are the terms of the *expansion* of $(q + p)^2$. Multiply $q + p$ by itself and you'll see what I mean:

$$
\begin{array}{r}
q + p \\
q + p
\end{array} \times
$$

$$
\begin{array}{r}
q^2 + qp \\
qp + p^2
\end{array} +
$$

$$
q^2 + 2qp + p^2
$$

0 Red   1 Red   2 Red | Each of these three terms stands for the probability
         1 White | of one particular combination.

This may give you a clue as to what will happen when we take a sample of 3*. Naturally, a sample of 3 marbles will give a greater variety of possible outcomes than will a sample of 2.

Extend your previous probability tree to show what can happen if we choose a third marble, and *list* all the possible colour sequences we might get.

CHECK your probability tree:

1st draw	2nd draw	3rd draw	Joint outcomes
		R	RRR
	R	W	RRW
R		R	RWR
	W	W	RWW
		R	WRR
	R	W	WRW
W		R	WWR
	W	W	WWW

Instead of the previous four joint outcomes (RR, RW, WR, and WW), we now have *eight*. Either R or W on the third choice may join with each of the four previous outcomes.

*Calculate $(q + p)^3$ if you can, and keep a note of your result.

So here we clearly have 8 permutations but not so many combinations. Just what have we got?
   (i)   How many different combinations are there? (e.g. 2 Red/1 White.)
   (ii) How many permutations give rise to each combination?
   (iii) What is the probability (in p's and q's) of each *permutation*?
   (iv) What is the total probability (in p's and q's) of each *combination*?

CHECK your results in the table below:

(a) There are 4 different combinations	(b) Either 1 or 3 permutations give rise to each combination	(c) Probability of each permutation.	(d)
↓ Type of result (Combination)	↓ Ways of arising (Permutation)	↓ Probability of each way	↓ Total probability of each combination
0 Red (and 3 White)	WWW	qqq	q^3
1 Red (and 2 White)	WWR WRW RWW	$qqp = q^2p$ $qpq = q^2p$ $pqq = q^2p$ $\Big\}$	$3q^2p$
2 Red (and 1 White)	RRW RWR WRR	$ppq = qp^2$ $pqp = qp^2$ $qpp = qp^2$ $\Big\}$	$3qp^2$
3 Red (and 0 White)	RRR	ppp	p^3

Notice that: $q^3 + 3q^2p + 3qp^2 + p^3 = (q + p)^3$
   So the left-hand column shows the only four possible results you could get from a sample of 3. The right-hand column shows the probability of each possible result.

   You know that, in the case of our marbles, $q = \dfrac{3}{4}$ and $p = \dfrac{1}{4}$ so:

Evaluate and *add up* the terms in the right-hand column:
$$q^3 + 3q^2p + 3qp^2 + p^3 = \underline{\ ?\ } + \underline{\ ?\ } + \underline{\ ?\ } + \underline{\ ?\ } = \underline{\ ?\ }$$

$$q^3 + 3q^2p + 3qp^2 + p^3 = \left(\frac{3}{4}\right)^3 + 3\left(\frac{3}{4}\right)^2\left(\frac{1}{4}\right) + 3\left(\frac{3}{4}\right)\left(\frac{1}{4}\right)^2 + \left(\frac{1}{4}\right)^3$$
$$= \frac{27}{64} + \frac{27}{64} + \frac{9}{64} + \frac{1}{64} = 1$$

So we have found the probability of each of the four possible combinations of Red and White we could get by taking a sample of 3. Notice that:

1 Each term stands for the	q^3	$3q^2p$	$3qp^2$	p^3
probability of one possible	0 Red	1 Red	2 Red	3 Red
combination	3 White	2 White	1 White	0 White

2 The number of combinations is *1 more* than the number of items in the sample.

3 The *indices* above the p's and q's tell us what the combination is, e.g. in $3q^2p$ we know that q^2p stands for 2 White and 1 Red.

4 The *coefficient* tells us how many ways that combination can arise, e.g. the 3 in $3q^2p$ says that q^2p can arise in three ways.

Make a note of this panel.

Suppose we take 4 marbles for our next sample:
(i) *How many* different combinations of Red and White could we have?
(ii) *What are* those combinations (e.g. 0 Red and 4 White, 3 Red and 1 White, etc.)?

With 4 marbles in the sample:
(i) There are 5 possible combinations. (One more than the number in the sample—because, although you could have n different numbers of Red marbles in the sample, you could *also* have 0 Red marbles.)
(ii) The five combinations are:

0 Red &	1 Red &	2 Red &	3 Red &	4 Red &
4 White	3 White	2 White	1 White	0 White

Now how can we represent these combinations in terms of p's and q's? Remember that the *indices* above the p and the q in each term tell us how many Reds are combined with how many Whites. So let us write the probability for each of the 5 types of combination mentioned above.

$\dfrac{0 \text{ Red \&}}{4 \text{ White}} = q^4$

$\dfrac{1 \text{ Red \&}}{3 \text{ White}} = q^3p \quad$ (p by itself $= p^1$, and $q = q^1$)

$\dfrac{2 \text{ Red \&}}{2 \text{ White}} = ?$

$\dfrac{3 \text{ Red \&}}{1 \text{ White}} = ?$

$\dfrac{4 \text{ Red \&}}{0 \text{ White}} = ?$

Complete the table to make up a term for each combination.

$0 \text{ Red} = q^4$

$\dfrac{1 \text{ Red \&}}{3 \text{ White}} = q^3p$

$$\begin{array}{l} \dfrac{2 \text{ Red \&}}{2 \text{ White}} = q^2 p^2 \\[1em] \dfrac{3 \text{ Red \&}}{1 \text{ White}} = qp^3 \\[1em] 4 \text{ Red} \quad = p^4 \end{array}$$

Now $q^2 p^2$ is NOT the *total* probability of getting 2 Red and 2 White in a sample of 4 marbles. It is merely the probability of getting one *particular* permutation (e.g. RWWR) of that combination. To find the total probability we need to know the number of possible permutations that could give rise to this combination—we need to know the *coefficient* that goes in front of $q^2 p^2$. Similarly for all the other terms:

$$\underline{?}\,q^4 + \underline{?}\,q^3 p + \underline{?}\,q^2 p^2 + \underline{?}\,qp^3 + \underline{?}\,p^4$$

How do we find these coefficients? Let's start with the easy ones:

What are the coefficients for q^4 and p^4? (How many ways can you get 4 Reds or 4 Whites?)

Since there's only one way you can get 4 Reds and one way you can get 4 Whites, the coefficient for q^4 and for p^4 is 1.

But we don't usually bother to write down a coefficient (or an index) of 1, so here is what we have got so far:

$$q^4 + \underline{?}\,q^3 p + \underline{?}\,q^2 p^2 + \underline{?}\,qp^3 + p^4$$

What about the other coefficients?

Here let us remember that the probabilities in a sample of 2 were given by the expansion of $(q + p)^2 = q^2 + 2qp + p^2$.

Similarly, the probabilities for the possible combinations in a sample of 3 were given by expanding $(q + p)^3 = q^3 + 3q^2 p + 3qp^2 + p^3$.

So you surely won't be surprised to know that the probabilities we are looking for now (with a sample of 4) turn out to be the expanded terms of $(q + p)^4 = (q + p)(q^3 + 3q^2 p + 3qp^2 + p^3)$.

That is:

$$\begin{array}{r} q^3 + 3q^2 p + 3qp^2 + \quad p^3 \\ q + \quad p \end{array} \times$$

$$\begin{array}{r} q^4 + 3q^3 p + 3q^2 p^2 + \quad qp^3 \\ q^3 p + 3q^2 p^2 + 3qp^3 + p^4 \\ \hline q^4 + 4q^3 p + 6q^2 p^2 + 4qp^3 + p^4 \end{array} +$$

See how the expansion of $(q + p)^4$ has put the coefficients on our probability terms—1, 4, 6, 4, 1.

Which of these terms represents the probability that our sample of 4 will contain 1 Red and 3 White marbles?

(a) $4q^3 p$?   or (b) $6q^2 p^2$?   or (c) $4qp^3$?

... $4q^3p$. The coefficient tells us that the combination of 1 Red and 3 White marbles can arise in 4 ways (i.e. RWWW, WRWW, WWRW, and WWWR). Similarly, the coefficients tell us that 3 Red and 1 White can also arise in 4 ways but 2 Red and 2 White can happen in 6 ways:

$$q^4 + 4q^3p + 6q^2p^2 + 4qp^3 + p^4$$

Now let's check these probabilities through to see that they behave as they should:

Combination   *Probability*

0 Red (4 White) $= q^4$ $= \left(\dfrac{3}{4}\right)^4$ $= \dfrac{81}{256}$

1 Red (3 White) $= 4q^3p$ $= 4\left(\dfrac{3}{4} \times \dfrac{3}{4} \times \dfrac{3}{4} \times \dfrac{1}{4}\right) = \dfrac{108}{256}$

2 Red (2 White) $= 6q^2p^2$ $= 6\left(\dfrac{3}{4} \times \dfrac{3}{4} \times \dfrac{1}{4} \times \dfrac{1}{4}\right) = \dfrac{54}{256}$

3 Red (1 White) $= 4qp^3$ $= \underline{\hspace{3cm}?}$ $= \dfrac{?}{256}$

4 Red (0 White) $= p^4$ $= \underline{\hspace{3cm}?}$ $= \dfrac{?}{256}$

(i) *Complete* the last two probability calculations.
(ii) What *should* be the *total* of all the probabilities on the right-hand side? Is it?

(i) $4qp^3 = 4\left(\dfrac{3}{4} \times \dfrac{1}{4} \times \dfrac{1}{4} \times \dfrac{1}{4}\right) = \dfrac{12}{256}$

(ii) $p^4 = \left(\dfrac{1}{4}\right)^4$ $\qquad\qquad = \dfrac{1}{256}$

$$\dfrac{81 + 108 + 54 + 12 + 1}{256} = \dfrac{256}{256} = 1$$

So here's what we've found so far:

Sample of 1.   Probability of 0 or 1 Reds $=$
$\qquad\qquad (q + p)^1 = q + p = 1$

Sample of 2.   Probability of 0 or 1 or 2 Reds $=$
$\qquad\qquad (q + p)^2 = q^2 + 2qp + p^2 = 1$

Sample of 3.   Probability of 0 or 1 or 2 or 3 Reds $=$
$\qquad\qquad (q + p)^3 = q^3 + 3q^2p + 3qp^2 + p^3 = 1$

Sample of 4.   Probability of 0 or 1 or 2 or 3 or 4 Reds $=$
$\qquad\qquad (q + p)^4 = q^4 + 4q^3p + 6q^2p^2 + 4qp^3 + p^4 = 1$

Clearly, if we took a sample of 5 marbles, the probabilities of choosing 0, 1, 2, 3, 4, and 5 Reds would be given by expanding the terms of

$$(q + p)^? \longleftarrow \text{---What is the index? (Don't worry if you don't}$$

actually know how to do the expansion).

... $(q + p)^5$

In general, we can say that:

If an item (e.g. a baby) can take either of two forms (e.g. boy or girl) whose probabilities are p and q, then the probabilities that a sample of n such items will contain 0, 1, 2, 3 ... $n - 3$, $n - 2$, $n - 1$, n boys are given by the successive terms of the expansion of $(q + p)^n$.

Make a note of this panel.

So, as you decided a moment ago, to find the probabilities of various numbers of Reds in a sample of 5, we need the expansion of $(q + p)^5$. Here are the first three terms of the expansion, together with the coefficients for the others:

$$(q + p)^5 = q^5 + 5q^4p + 10q^3p^2 + 10\underline{\ ?\ } + 5\underline{\ ?\ } + \underline{\ ?\ }$$

Can you *complete* the three final terms?

$$(q + p)^5 = q^5 + 5q^4p + 10q^3p^2 + 10q^2p^3 + 5qp^4 + p^5$$

The Term and its Coefficient

Notice that, as in all the expansions so far:

1   The *coefficients* are *symmetrical*. The end-terms both have a coefficient of 1 since there is only one way of getting all Reds or all Whites. The more equal the numbers of each kind in the combination, the larger the coefficient—the more ways there are of getting that combination.

2   The *p index* increases in size from left to right as the number of Reds in a sample increases. But as the number of Reds rises, the number of Whites must fall; so the *q* index decreases from left to right. (For each term, the total of the two indices is always equal to n.)

Suppose we now take a sample of 6 marbles:

(i)   We can find the probabilities of 0, 1, 2, etc. Reds by expanding the terms of $\underline{\ ?\ }$.

(ii)   How many combinations would this give us?

(iii)   Write out the combinations in p's and q's (starting with q^6 and q^5p), then I'll give you the coefficients.

(i)   We can find the probabilities of 0, 1, 2, etc. Reds by expanding the terms of $(q + p)^6$.

(ii)   There will be *seven* combinations (that is, $n + 1$).

(iii)   These are the combinations: q^6, q^5p, q^4p^2, q^3p^3, q^2p^4, qp^5 and p^6. And by expanding $(q + p)^6$ we could write in the coefficients to show how many ways each combination could arise:

$$q^6 + 6q^5p + 15q^4p^2 + 20q^3p^3 + 15q^2p^4 + 6qp^5 + p^6$$

Check once again that the coefficients are symmetrical and that one index falls as the other one rises.

(i) The term $15q^2p^4$ represents the probability that our sample of 6 marbles contains _?_ Reds and _?_ Whites.

(ii) The term $6qp^5$ represents ___?___

(iii) What terms would you need to expand to find the probability of getting 8 Reds and 2 Whites? $(q + p)^?$

(i) 4 Reds and 2 Whites.

(ii) $6qp^5 = $ probability of getting 5 Reds and 1 White (in a sample of 6).

(iii) To find the probability of getting 8 Reds and 2 Whites we would need to expand the terms of $(q + p)^{10}$ for we would be dealing with a sample of 10.

Fortunately, you have no need to carry out these expansions. There are easier ways of finding the coefficients for your probability terms. Look at our results so far:

No. in sample (n)	Expansion of $(q + p)^n$	No. of combinations
1	$q + p$	2
2	$q^2 + 2qp + p^2$	3
3	$q^3 + 3q^2p + 3qp^2 + p^3$	4
4	$q^4 + 4q^3p + 6q^2p^2 + 4qp^3 + p^4$	5
5	$q^5 + 5q^4p + 10q^3p^2 + 10q^2p^3 + 5qp^4 + p^5$	6
6	$q^6 + 6q^5p + 15q^4p^2 + 20q^3p^3 + 15q^2p^4 + 6qp^5 + p^6$	7

We've already noticed that the coefficients are symmetrical. Is there anything else you notice—about the *size* of the coefficients? (Pick on a coefficient somewhere near the middle of a row and compare it with those in the row above it.)

Well, you may have noticed the connection between the size of the coefficients already. In a moment, we'll check.

Coefficients from Pascal's Triangle

In fact, the relationship between the coefficients in the expansions of $(q + p)^n$ was pointed out many years ago by the mathematician Blaise Pascal. Nowadays, this arrangement of numbers is called 'Pascal's Triangle'.

No. in sample (n)	Coefficients in expansion of $(q + p)^n$						
1				1	1		
2			1	2	1		
3		1	3	3	1		
4	1	4	6	4	1		
5	1	5	10	10	5	1	
6	1	6	15	20	15	6	1

What Pascal spotted was the rule by which rows can be added on to the base of the triangle—so we could go on to write out the coefficients for the expanded terms of $(q + p)^7$, $(q + p)^8$, and so on.

Did you see the rule? Look at any coefficient in the triangle—then look at the coefficient to the left and to the right of it in the row above. What is the connection?

You probably noticed that each coefficient in the triangle is found by ADDING TOGETHER the coefficients lying to either side of it in the row above.

No. in sample (n)	Coefficients in expansion of $(q + p)^n$							
1				1	1			
2			1	2	1			
3		1	3	3	1			
4	1	4	6	4	1			
5	1	5	10	10	5	1		
6	1	6	15	20	15	6	1	
7								

e.g.
$1 + 2 = 3$
$1 + 0 = 1$
$1 + 5 = 6$
$10 + 5 = 15$

Suppose we now add another row to the triangle—to show the coefficients of the expansion when $n = 7$. When we take a sample of 7, we can get any of 8 combinations.

Write out the 8 coefficients for the combinations.

We get the 8 coefficients for the combinations possible when $n = 7$ by adding another row to Pascal's Triangle:

No. in sample (n)	Coefficients in expansion of $(q + p)^n$								
1					1	1			
2				1	2	1			
3			1	3	3	1			
4		1	4	6	4	1			
5	1	5	10	10	5	1			
6	1	6	15	20	15	6	1		
7	1	7	21	35	35	21	7	1	

So Pascal's Triangle gives us a way of finding the total probability of any binomial combinations. All we do is:

1   Write down the $n + 1$ combinations in terms of p and q.
2   Add rows to Pascal's Triangle until we get the $n + 1$ coefficients.
3   Evaluate the combinations we are interested in.

For example, suppose we want to find the probabilities of the various combinations of Reds and Whites if we take a sample of $n = 7$ marbles. All we need do is write out the combinations from q^7 to p^7, attach the coefficients from the last row you added to the triangle, and then evaluate the completed terms. Try it:

In terms of p and q, what is the probability that a sample of 7 marbles will contain 4 Reds and 3 Whites?

$35q^3p^4$ is the probability that the sample of 7 contains 4 Reds.

The indices in q^3p^4 stand for '3 White, 4 Red' and the coefficient of 35 indicates the number of ways that combination can arise. The coefficients for all the terms of $(q + p)^7$ were easy to see in the row you added to Pascal's Triangle. Thus:

$(q + p)^7 = q^7 + 7q^6p + 21q^5p^2 + 35q^4p^3 + 35q^3p^4 + 21q^2p^5 + 7qp^6 + p^7$

Similarly,
the probability of 2 Reds is $21q^5p^2$
the probability of 6 Reds is $7qp^6$
the probability of 1 Red is $7q^6p$ and so on.

It would now be easy to go on to stage 3 (by recalling that $p = \frac{1}{4}$ and $q = \frac{3}{4}$) and so evaluate each probability combination. However, let us first have some more practice with stages 1 and 2—writing down the $n + 1$ terms from q^n to p^n and getting the coefficients from Pascal's Triangle.

Suppose we want a sample of 5 marbles. *Write out* all the possible combinations in terms of p and q.

You should have written $n + 1 = 6$ combinations:
$$q^5, \quad q^4p, \quad q^3p^2, \quad q^2p^3, \quad qp^4, \quad p^5.$$

That was stage 1; now for stage 2.

We need to know how many ways each combination can arise—what are the coefficients? So we'll need to add rows to Pascal's Triangle until we get to the row for $n - 1$.

Can you write out Pascal's Triangle using the rule you discovered a few pages back? I'll give you a start:

$$1 \qquad \qquad 1$$
$$? \qquad \qquad ? \qquad \qquad ?$$

Write out the triangle until you get to the row you need.

Here is Pascal's Triangle down to $n = 5$:

$$
\begin{array}{ccccccccccc}
 & & & & 1 & & 1 & & & & \\
 & & & 1 & & 2 & & 1 & & & \\
 & & 1 & & 3 & & 3 & & 1 & & \\
 & 1 & & 4 & & 6 & & 4 & & 1 & \\
1 & & 5 & & 10 & & 10 & & 5 & & 1
\end{array}
$$

Thus the complete probability terms for $n = 5$ are:
$$q^5 \quad 5q^4p \quad 10q^3p^2 \quad 10q^2p^3 \quad 5qp^4 \quad p^5$$

So we've got the combinations and we know their coefficients. Let's go on to stage 3 this time and evaluate the probabilities.

In our huge box of marbles, remember, $\frac{1}{4}$ are Red and $\frac{3}{4}$ are White, so:

Probability (0 Red) $= q^5 = \left(\frac{3}{4}\right)^5 \qquad = \dfrac{243}{1024}$

Probability (1 Red) $= 5q^4p = 5\left(\frac{3}{4}\right)^4\left(\frac{1}{4}\right) \qquad = \dfrac{405}{1024}$

Probability (2 Red) $= 10q^3p^2 = 10\left(\frac{3}{4}\right)^3\left(\frac{1}{4}\right)^2 = \dfrac{270}{1024}$

You try the next one:

What is the probability of choosing 3 Reds?

According to the combinations you wrote out:
$$q^5 \quad 5q^4p \quad 10q^3p^2 \quad \underline{10q^2p^3} \quad 5qp^4 \quad p^5$$
the probability of 3 Reds is $10q^2p^3$, and since $p = \frac{1}{4}$ and $q = \frac{3}{4}$

$$10q^2p^3 = 10\left(\frac{3}{4}\right)^2\left(\frac{1}{4}\right)^3 = 10\left(\frac{3 \cdot 3 \cdot 1 \cdot 1 \cdot 1}{1024}\right)$$

$$\text{Probability (3 Reds)} = \frac{90}{1024}$$

So here are the probabilities we've calculated so far.

When $n = 5$, and $p = \frac{1}{4}$ and $q = \frac{3}{4}$

Probability (0 Red) $= q^5 \qquad\quad = \dfrac{243}{1024}$

Probability (1 Red) $= 5q^4p \qquad = \dfrac{405}{1024}$

Probability (2 Red) $= 10q^3p^2 = \dfrac{270}{1024}$

Probability (3 Red) $= 10q^2p^3 = \dfrac{90}{1024}$

Probability (4 Red) $= \qquad\qquad\qquad ?$

Probability (5 Red) $= \qquad\qquad\qquad ?$

(i) *Calculate* the two remaining probabilities.

(ii) *Add* all six probabilities together as a check on your arithmetic.

(i) The two remaining probabilities are:
$$\text{Probability (1 Red)} = 5qp^4 = 5\left(\frac{3}{4}\right)\left(\frac{1}{4}\right)^4 = \frac{15}{1024}$$

$$\text{Probability (0 Red)} = q^5 \quad = \left(\frac{1}{4}\right)^5 \quad = \frac{1}{1024}$$

(ii) And, as we would expect, since $(q + p)^n = 1$, the six probabilities total to 1:
$$\frac{243 + 405 + 270 + 90 + 15 + 1}{1024} = \frac{1024}{1024} = 1$$

(In a sample of 5 we are *certain* to get either 0 or 1 or 2 or 3 or 4 or 5 Red marbles.)

Having worked out these probabilities we could put them together to find yet others. For instance, the probability that a sample of 5 marbles will contain *either* 5 *or* 4 Reds is
$$\frac{1}{1024} + \frac{15}{1024} = \frac{16}{1024}$$

What is the probability that a sample of 5 will contain
  (i)   more Reds than Whites?
  (ii) more Whites than Reds?

(i) To calculate the probability of getting more Reds than Whites we must add together the probabilities of getting 3, 4, and 5 Reds:
$$10q^2p^3 + 5qp^4 + p^5 = \frac{90 + 15 + 1}{1024} = \frac{106}{1024}$$

(ii) To calculate the probability of more Whites than Reds, we add the probabilities of 0, 1, and 2 Reds:
$$q^5 + 5q^4p + 10q^3p^2 = \frac{243 + 405 + 270}{1024} = \frac{918}{1024}$$

So far we've done all our sampling from a box of marbles, $\frac{1}{4}$ of which are Red and the rest White. We have worked out the probabilities of getting various proportions of Red and White in samples of various size, e.g. in a sample of $n = 5$:

0 Red	1 Red	2 Red	3 Red	4 Red	5 Red
$\frac{243}{1024}$	$\frac{405}{1024}$	$\frac{270}{1024}$	$\frac{90}{1024}$	$\frac{15}{1024}$	$\frac{1}{1024}$

Suppose we added more marbles to the box, so that $\frac{1}{3}$ of the marbles are now Reds. Would we still have the same chances of getting 0, 1, 2, 3, 4, and 5 Reds in a sample of 5? Yes or No?

No, we would not have the same chances.

If the proportions in the box change, then the probabilities of getting 0, 1, 2, 3, 4, and 5 Reds in a sample of 5 will also change.

These probabilities are defined in terms of p and q.

0 Red	1 Red	2 Red	3 Red	4 Red	5 Red
q^5	$5q^4p$	$10q^3p^2$	$10q^2p^3$	$5qp^4$	p^5

and if the values of p and q change from $p = \dfrac{1}{4}$ and $q = \dfrac{3}{4}$ to $p = \dfrac{1}{3}$ and $q = \dfrac{2}{3}$, then clearly the probability of each combination is also changed.

(i) Calculate the probability of each combination (and check your arithmetic by finding the total).

(ii) What is the probability of choosing five marbles that are all the same colour?

(i) The probabilities are:

$$0 \text{ Red: } q^5 = \left(\frac{2}{3}\right)^5 = \frac{32}{243}$$

$$1 \text{ Red: } 5q^4p = 5\left(\frac{2}{3}\right)^4\left(\frac{1}{3}\right) = \frac{80}{243}$$

$$2 \text{ Red: } 10q^3p^2 = 10\left(\frac{2}{3}\right)^3\left(\frac{1}{3}\right)^2 = \frac{80}{243}$$

$$3 \text{ Red: } 10q^2p^3 = 10\left(\frac{2}{3}\right)^2\left(\frac{1}{3}\right)^3 = \frac{40}{243}$$

$$4 \text{ Red: } 5qp^4 = 5\left(\frac{2}{3}\right)\left(\frac{1}{3}\right)^4 = \frac{10}{243}$$

$$5 \text{ Red: } p^5 = \left(\frac{1}{3}\right)^5 = \frac{1}{243}$$

(And $\dfrac{32 + 80 + 80 + 40 + 10 + 1}{243} = \dfrac{243}{243} = 1$.)

(ii) For all the marbles to be the same colour you must choose *either*

$$5 \text{ White } or \text{ 5 Red: } q^5 + p^5 = \frac{32 + 1}{243} = \frac{33}{243}$$

Suppose we now change the proportions slightly so that 30% of the marbles are red and 70% white. This time we'll take a sample of 4.

(i) Write out the $n + 1$ possible combinations—from q^n to p^n.

(ii) Use Pascal's Triangle to get the coefficients.

(iii) Calculate the probability of each combination.

(i) The possible combinations in a sample of $n = 4$ are:
$$q^4, \quad q^3p, \quad q^2p^2, \quad qp^3, \quad p^4$$

(ii) And the coefficients come from the 4th row of Pascal's Triangle;

```
        1       1
      1    2    1
    1    3    3    1
  1    4    6    4    1
```

giving $q^4 \quad 4q^3p \quad 6q^2p^2 \quad 4qp^3 \quad p^4$

(iii) Probability (0 Red) $= q^4 \quad = \left(\dfrac{7}{10}\right)^4 \quad = \dfrac{2401}{10\,000}$

Probability (1 Red) $= 4q^3p \quad = 4\left(\dfrac{7}{10}\right)^3\left(\dfrac{3}{10}\right) = \dfrac{4116}{10\,000}$

Probability (2 Red) $= 6q^2p^2 = 6\left(\dfrac{7}{10}\right)^2\left(\dfrac{3}{10}\right)^2 = \dfrac{2646}{10\,000}$

Probability (3 Red) $= 4qp^3 \quad = 4\left(\dfrac{7}{10}\right)\left(\dfrac{3}{10}\right)^3 = \dfrac{756}{10\,000}$

Probability (4 Red) $= p^4 \quad = \left(\dfrac{3}{10}\right)^4 \quad = \dfrac{81}{10\,000}$

And: $\dfrac{2401 + 4116 + 2646 + 756 + 81}{10\,000} = \dfrac{10\,000}{10\,000} = 1.$

Let's change the proportions again:

Suppose we add yet more marbles to the box until we have half red marbles and half white.

(i) Now what are the probabilities of getting 0, 1, 2, 3, and 4 Reds in a sample of 4?

(ii) Compare the probabilities of getting more Reds than Whites and more Whites than Reds.

(i) Probability (0 Red) $= q^4 \quad = \left(\dfrac{1}{2}\right)^4 \quad = \dfrac{1}{16}$

Probability (1 Red) $= 4q^3p \quad = 4\left(\dfrac{1}{2}\right)^3\left(\dfrac{1}{2}\right) \quad = \dfrac{4}{16}$

Probability (2 Red) $= 6q^2p^2 = 6\left(\dfrac{1}{2}\right)^2\left(\dfrac{1}{2}\right)^2 = \dfrac{6}{16}$

Probability (3 Red) $= 4qp^3 \quad = 4\left(\dfrac{1}{2}\right)\left(\dfrac{1}{2}\right)^3 \quad = \dfrac{4}{16}$

Probability (4 Red) $= p^4 \quad = \left(\dfrac{1}{2}\right)^4 \quad = \dfrac{1}{16}$

And $\dfrac{1 + 4 + 6 + 4 + 1}{16} = \dfrac{16}{16} = 1.$

Notice that this is a special case: because $p = q = \dfrac{1}{2}$, the probabilities are symmetrical.

(ii) Because $p = q = \dfrac{1}{2}$, the chance of getting more Reds than Whites, $\dfrac{1}{16} + \dfrac{4}{16} = \dfrac{5}{16}$, is exactly the same as the chance of getting more Whites than Reds, $\dfrac{4}{16} + \dfrac{1}{16} = \dfrac{5}{16}.$

Are you getting just a little tired of marbles by now? Let's have a change of scenery:

If 4 pennies are tossed simultaneously, what is the probability you'll get 2 heads and 2 tails? (You don't need to do any working at all for this one.)

The combination you need was q^2p^2 and there was no need to work out the coefficient in Pascal's Triangle for $n = 4$ because you'd just done it for the marbles. $6q^2p^2$ was the total probability of getting two of each kind in a sample of 4. And since $p = q = \frac{1}{2}$,

$$6q^2p^2 = 6\left(\frac{1}{2}\right)^2\left(\frac{1}{2}\right)^2 = \frac{6}{16}$$

is the same result as you got with the marbles.

Try another.
The probability of a rabbit from a litter being black is $\frac{1}{4}$. What is the probability of getting 3 black rabbits in a litter of 5?

The probability of getting 3 black rabbits in a litter of 5 is $\frac{90}{1024}$.

We were interested in the combination q^2p^3 and the $n = $ 5th row in Pascal's Triangle (1   5   10   10   5   1) gave us the coefficient for the expanded terms of $(q + p)^5$:

$$q^5 \quad 5q^4p \quad 10q^3p^2 \quad 10q^2p^3 \quad 5qp^4 \quad p^5$$

And, with $p = \frac{1}{4}$, and $q = \frac{3}{4}$, $10q^2p^3 = 10\left(\frac{3}{4}\right)^2\left(\frac{1}{4}\right)^3 = \frac{90}{1024}$

In a certain factory, workmen have a 20% chance of suffering at some time from an industrial disease. What is the probability that exactly half of the six new men employed today will eventually catch the disease?

We have a sample of 6 and we need the coefficient for the probability term q^3p^3. From the 6th row of Pascal's Triangle,

$$
\begin{array}{ccccccccccc}
 & & & & & 1 & & 1 & & & \\
 & & & & 1 & & 2 & & 1 & & \\
 & & & 1 & & 3 & & 3 & & 1 & \\
 & & 1 & & 4 & & 6 & & 4 & & 1 \\
 & 1 & & 5 & & 10 & & 10 & & 5 & & 1 \\
1 & & 6 & & 15 & & 20 & & 15 & & 6 & & 1
\end{array}
$$

we see that the complete term is

$$q^6 \quad 6q^5p \quad 15q^4p^2 \quad \underline{20q^3p^3} \quad 15q^2p^4 \quad 6qp^5 \quad p^6$$

And, since $p = 20\% = \frac{1}{5}$ and $q = \frac{4}{5}$: $20q^3p^3 = 20\left(\frac{4}{5}\right)^3\left(\frac{1}{5}\right)^3 = \frac{256}{3125}$

On a particular political issue, opinion polls suggest that 75% of the voters support the president. If TV news interviews 4 people chosen at random, what is the probability that exactly 3 of them will support the president?

In this case we need the coefficient for the terms qp^3 and we get it from the $n = $ 4th row of Pascal's Triangle.

$$\begin{array}{ccccccc} & & & 1 & & 1 & \\ & & 1 & & 2 & & 1 \\ & 1 & & 3 & & 3 & & 1 \\ 1 & & ④ & & 6 & & 4 & & 1 \end{array}$$

Thus we see that there are 4 permutations of 3 supporters and 1 opponent and, since $p = \dfrac{3}{4}$ and $q = \dfrac{1}{4}$, $4qp^3 = 4\left(\dfrac{1}{4}\right)\left(\dfrac{3}{4}\right)^3 = \dfrac{27}{64}$.

15% of the cars made by a certain firm have a fault so serious that customers need to take a car back to the dealer within a week of purchase. If a dealer sells three cars in one day, what is the probability that all three customers will be back to complain within a week? Two of them? One of them? None of them?

We need the coefficients for the combinations q^3, q^2p, qp^2 and p^3. Since $n = 3$ we get them from the 3rd row of Pascal's Triangle.

$$\begin{array}{ccccccc} & & & 1 & & 1 & \\ & & 1 & & 2 & & 1 \\ & 1 & & 3 & & 3 & & 1 \end{array}$$

So we need to evaluate the terms
$$q^3 \quad 3q^2p \quad 3qp^2 \quad p^3$$
where $p = \dfrac{15}{100} = \dfrac{3}{20}$ and $q = \dfrac{17}{20}$.

Probability (0 complaints) $= q^3 = \left(\dfrac{17}{20}\right)^3 = \dfrac{4913}{8000}$

Probability (1 complaints) $= 3q^2p = 3\left(\dfrac{17}{20}\right)^2\left(\dfrac{3}{20}\right) = \dfrac{2601}{8000}$

Probability (2 complaints) $= 3qp^2 = 3\left(\dfrac{17}{20}\right)\left(\dfrac{3}{20}\right)^2 = \dfrac{459}{8000}$

Probability (3 complaints) $= p^3 = \left(\dfrac{3}{20}\right)^3 = \dfrac{27}{1000}$

And you should have checked your arithmetic by seeing that the probabilities add up to 1:
$$\dfrac{4913 + 2601 + 459 + 27}{8000} = \dfrac{8000}{8000} = 1$$

One more to finish this section:
A factory has five telephone lines and the probability that any one of them is

engaged at a given moment during the day is $\frac{2}{7}$. Calculate the probability that at least three of the lines will be engaged if I try to contact the factory now.

$n = 5$ and the six possible combinations are
$$q^5 \quad q^4p \quad q^3p^2 \quad q^2p^3 \quad qp^4 \quad p^5$$
And from the 5th row of Pascal's Triangle we get the coefficients:
$$q^5 \quad 5q^4p \quad 10q^3p^2 \quad 10q^2p^3 \quad 5qp^4 \quad p^5$$
Since we are looking for the probability that at least 3 (that is, 3, 4, or 5) lines are engaged, we need to evaluate:

$$
\begin{aligned}
10q^2p^3 + 5qp^4 + p^5 &= 10\left(\frac{5}{7}\right)^2\left(\frac{2}{7}\right)^3 + 5\left(\frac{5}{7}\right)\left(\frac{2}{7}\right)^4 + \left(\frac{2}{7}\right)^5 \\
&= \frac{2000}{16\,807} + \frac{400}{16\,807} + \frac{32}{16\,807} \\
&= \frac{2432}{16\,807} = 0.145
\end{aligned}
$$

(About once in every seven calls you make, you'll find 3 or more lines engaged.)

Review

So you now have a very good way of calculating *binomial* probabilities—where an item can take either of two forms (with probabilities p and q) and you want to know the probability that a sample of so many items will contain a given number of each form.
Stage 1   Write down the required combination in terms of p and q.
Stage 2   Attach the coefficients from the nth row of Pascal's Triangle.
Stage 3   Evaluate the required probability term.

Coefficients from Combinations

In the previous section you learned how to evaluate binomial probabilities using Pascal's Triangle. First you write down the required 'mixture' in terms of p and q (e.g. q^3p^2). Then you attach the appropriate *coefficient* which you obtain from the nth row of Pascal's Triangle (e.g. $10q^3p^2$).

Now this method of finding the coefficient is fine when you are dealing with small samples, say 12 or less. But suppose, for instance, that you wanted to know the probability that in a church congregation of 50 people, 5 are men and the rest women. Here you could write the probability term as $q^{45}p^5$ (where p is the probability of a man and q of a woman). But how many ways could this particular 'mixture' arise? How, in other words, would you find the *coefficient*?

You would have to write out (*how many?*) rows of Pascal's Triangle to reach the row you need—and that row would certain (*how many?*) coefficients.

You would have to write out 50 rows of Pascal's Triangle to reach the row you need—and that row would contain 51 coefficients.

Clearly, the writing of these rows would take a long time, especially when you consider the size of the numbers you would be dealing with. The $n = 50$th row would begin:

$$1 \quad 50 \quad 1225 \quad 19\,600 \quad 230\,300 \quad 2\,118\,760 \text{ etc.,}$$

You would be quite likely to make errors in your additions, and it would be very tedious to go to all this trouble for just the one coefficient you need ($2\,118\,760 \, q^{45}p^5$).

Fortunately, there is a simpler method of deciding on the coefficient. Suppose we are taking a sample of 5 from a large number of electric light bulbs, some of which are faulty, and we want to know the probability that our sample will contain only two good bulbs.

First of all, if p is the probability of a good bulb and q of a faulty one, how would you write the probability (in terms of p and q) for 2 good and 3 faulty?

---

q^3p^2 is the probability term for 2 good bulbs and 3 faulty.

Now what is the total probability of getting just 2 good bulbs in a sample of 5? We know that q^3p^2 is the probability that the 2 good bulbs will occur in any one particular position (e.g. first and last, or second and third). But the 2 good bulbs can be *any* pair among the 5. So how many pairs are possible? (Multiply q^3p^2 by this coefficient and we'll have the probability we are looking for.)

Once more, the idea of *combinations* comes to our aid. For all we are really asking is: how many ways can we choose 2 positions out of the 5 for the 2 good bulbs to appear in? That is, how many different *combinations of* 2 can we get *from 5*?

Do you remember how to calculate this number of combinations—5C2? (If you need a hint, look at the footnote.)*

$$n\mathrm{C}r = \frac{n!}{(n-r)!r!} = \frac{5!}{3!2!} = 10.$$

And you can check this result from Pascal's Triangle:

```
            1       1
        1       2       1
      1     3       3     1
    1     4     6     4     1
  1     5    10    10     5     1
```

So we see there are 10 ways of getting 2 positions out of 5 occupied by good bulbs (the other 3 positions being occupied by faulty bulbs).

*Remember: $n\mathrm{C}r = \dfrac{n!}{(n-r)!r!}$

And here *are* the 10 possible positions for the 2 good bulbs:

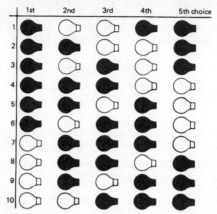

Clearly then, the coefficient for the term q^3p^2 (3 dud, 2 good) is 10—there are 10 ways this can happen.

So the probability of getting just 2 good bulbs in a sample of 5 is _?_ q^3p^2.

So the probability of getting just 2 good bulbs in a sample of 5 is $\underline{10}q^3p^2$.

Calculating the Coefficient

Instead of using Pascal's Triangle to find this coefficient, you simply calculated the number of ways of choosing 2 positions out of 5—you calculated the number of combinations among 5 positions taken 2 at a time: 5C2.

Now suppose we wanted to known the probability of getting 7 good bulbs (and 3 duds) in a sample of 10. How would you calculate the coefficient for q^3p^7? (Remember, $nCr = \dfrac{n!}{(n-r)!r!}$.)

$\dfrac{10!}{3!7!}$ is the coefficient for q^3p^7.

Here we need to choose 7 positions out of 10 for the good bulbs to occupy (the remaining 3 automatically being faulty). The number of combinations of 7 from 10 is

$10C7 = \dfrac{10!}{3!7!} = 120$ ways of getting 7 good bulbs (and 3 faulty) in a sample of 10.

So far, then, we've found it quite easy to calculate the coefficients for these two probability terms:

$$\frac{5!}{3!2!}q^3p^2 \quad \text{and} \quad \frac{10!}{3!7!}q^3p^7$$

without using Pascal's Triangle.

Try another:

What is the probability of choosing 4 good bulbs in a sample of 12? (Write it in the form shown above.)

$\frac{12!}{8!4!}q^8p^4$ is the probability of choosing 4 good bulbs in a sample of 12. The coefficient $\frac{12!}{8!4!}$ is simply the number of ways combining 4 good bulbs with 8 duds.

So we've calculated three coefficients so far:

$$\frac{5!}{3!2!}q^3p^2 \qquad \frac{10!}{3!7!}q^3p^7 \qquad \frac{12!}{8!4!}q^8p^4$$

Now try these.

What is the probability of choosing...

   (i)   3 good bulbs in a sample of 10?   _____q^7p^3

   (ii) 8 good bulbs in a sample of 12?   _____q^4p^8

   (iii) *r* good bulbs in a sample of *n* bulbs? _____$q^{n-r}p^r$

(i) $\frac{10!}{7!3!}q^7p^3$

(ii) $\frac{12!}{4!8!}q^4p^8$

(iii) $\frac{n!}{(n-r)!r!}\ q^{n-r}p^r$

In each case you found the coefficient by calculating the number of combinations available from *n* items (positions) taken *r* at a time. So, as we've seen in the last few pages:

> If you want to calculate the probability of getting *r* of one kind of thing in a sample of *n* things, the coefficient for the probability term $q^{n-r}p^r$ is *n*C*r*.
> That is, Probability $= (nCr)q^{n-r}p^r$
> $$= \frac{n!}{(n-r)!r!}\ q^{n-r}p^r$$

Make a note of this panel.

Now, to return to our first example, what is the probability of finding 5 men (and 45 women) in a congregation of 50? (Don't multiply it right out.)

The probability of getting 5 men and 45 women in a sample of 50 is $\dfrac{50!}{45!5!}q^{45}p^5$.

So now we have a quick way of calculating the coefficient for any individual probability term $q^{n-r}p^r$. Just remember that the coefficient is equal to the number of ways of combining n items taken r at a time. (It is *also* equal to the number of ways of permuting n items of which r are alike of one kind and $n - r$ are alike of another kind.)

Another example: the probability of getting 3 red and 4 white marbles in a sample of 7 is

$$(7C3)q^4p^3 = \frac{7!}{4!3!}q^4p^3 = \frac{7\cdot6\cdot5\cdot4!}{4!3\cdot2\cdot1} = 35q^4p^3$$

Try another example you worked out in the previous section:

If four pennies are tossed together, what is the probability of getting 2 heads and 2 tails? (Work it right out.)

The probability of getting 2 heads and 2 tails is:

$$(4C2)q^2p^2 = \frac{4!}{2!2!}q^2p^2 = \frac{4\cdot3\cdot2!}{2!2\cdot1}q^2p^2 = 6q^2p^2 \text{ as before.}$$

And, since $p = \dfrac{1}{2} = q$:

$$6q^2p^2 = 6\left(\frac{1\cdot1\cdot1\cdot1}{2\cdot2\cdot2\cdot2}\right) = \frac{6}{16}$$

Try the combinations method on this problem from the previous section:

On a particular political issue, opinion polls suggest that 75% of the voters support the president. If TV news interviews 4 people chosen at random, what is the probability that exactly 3 of them will support the president?

$$(4C3)qp^3 = \frac{4!}{3!1!}qp^3 = \frac{4\cdot3!}{3!1!}qp^3 = 4qp^3$$

(And since $p = \dfrac{3}{4}$ and $q = \dfrac{1}{4}$)

$$4qp^3 = 4\left(\frac{1}{4}\right)\left(\frac{3}{4}\right)^3$$

$$= \frac{27}{64} \text{ as before}$$

What does this probability of $\dfrac{27}{64}$ mean? It means that if we took 64 samples of 4 people we should expect to find 75% of the people in the sample supporting the president in about (*how many?*) of these samples.

... in about *27* of those samples. (That is we'd expect to find, in the long run, three out of a sample of four people supporting the president in about $\frac{27}{64} = 42\%$ of such samples.)

Now check through this new problem using the combinations method: A quarterback has a $\frac{1}{10}$ chance of completing a pass with each ball he throws. What is the probability he will complete exactly three passes in his next 6 balls?

$$
\begin{aligned}
\text{Probability of} \atop \text{3 completions in 6 passes} &= \frac{n!}{(n-r)!r!} \, q^{n-r}p^r \\
&= \frac{6!}{3!3!}\left(\frac{9}{10}\right)^3\left(\frac{1}{10}\right)^3 \\
&= 20\left(\frac{9}{10}\right)^3\left(\frac{1}{10}\right)^3 \\
&= 20\left(\frac{1}{1000}\right)\left(\frac{729}{1000}\right) \\
&= \frac{729}{50\,000}
\end{aligned}
$$

Here is another way of looking at this probability. In 50 000 passes a quarterback would expect to get 3 completions out of 6 on (*how many?*) occasions.

... a quarterback would expect to get 3 completions out of 6 on *729* occasions.

A survey is being taken of Army families with 3 children. If we assume that boys and girls are equally likely, what is the probability that such a family will have exactly 3, 2, 1, or 0 boys?

Probability (3 boys) =
$$(3\text{C}3)p^3 \;=\; \frac{3!}{0!3!}p^3 \;=\; p^3 \;=\; \left(\frac{1}{2}\right)^3 \;=\; \frac{1}{8}$$

Probability (2 boys) =
$$(3\text{C}2)qp^2 \;=\; \frac{3!}{1!2!}qp^2 \;=\; 3qp^2 \;=\; 3\left(\frac{1}{2}\right)\left(\frac{1}{2}\right)^2 \;=\; \frac{3}{8}$$

Probability (1 boy) =
$$(3\text{C}1)q^2p \;=\; \frac{3!}{2!1!}q^2p \;=\; 3q^2p \;=\; 3\left(\frac{1}{2}\right)^2\left(\frac{1}{2}\right) \;=\; \frac{3}{8}$$

Probability (0 boys) =
$$(3\text{C}0)q^3 \;=\; \frac{3!}{3!0!}q^3 \;=\; q^3 \;=\; \left(\frac{1}{2}\right)^3 \;=\; \frac{1}{8}$$

Each of the probabilities above, $\frac{3}{8}$ for example, is the probability that any given family will have a certain number of boys. But of course we can use these same probabilities to estimate the *relative frequency* with which each number of boys might occur among any number of such families.

Thus, among every *eight* such families you would expect to find, *on average*, that

(i)   _?_ of them had 3 boys.
(ii) _?_ of them had 2 boys.
(iii) _?_ of them had 1 boy.
(iv) _?_ of them had 0 boys.

Thus, among every *eight* such families you would expect to find, *on average*, that

(i)   *one* of them had 3 boys.
(ii) *three* of them had 2 boys.
(iii) *three* of them had 1 boy.
(iv) *one* of them had 0 boys.

Expected Frequency Distributions

What you have worked out above is usually called an *expected frequency distribution*. We can illustrate it with a diagram called a histogram. It shows the frequency with which each possible number of boys (0, 1, 2, or 3) can be expected, *in theory*, among eight 3-child families.

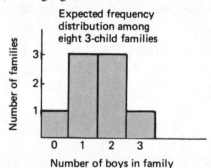

Now it is important not to be misled by the word 'expected' here. Let's make sure we know just what it means. For instance, does it mean that, in *each and every* possible set of eight 3-child families, we'd find exactly one with 0 boys, three with 1 boy, three with 2 boys, and one with 3 boys? Why or why not?

No, you would not find exactly these proportions among each and every set of eight you happened to take. You might easily get a set of eight 3-child families with all boys or all girls, for instance. (Just as a tossed coin may come down heads ten times in a row, even though the theoretical expectation might be 5 heads and 5 tails.)

So the distribution above is not to be expected for any particular set of eight. Rather it is expected *on average* (*in the long run*) across all possible sets of eight 3-child families. It is a *theoretical* expectation.

Here now is a binomial probability distribution you worked out in the last section. It shows the probabilities of getting various numbers of red marbles in a sample of 4 when $p = \dfrac{1}{2} = q$:

0 Red	1 Red	2 Red	3 Red	4 Red
$\dfrac{1}{16}$	$\dfrac{4}{16}$	$\dfrac{6}{16}$	$\dfrac{4}{16}$	$\dfrac{1}{16}$

Suppose you drew exactly 16 samples of 4. How many samples would you 'expect' (theoretically) to contain 0, 1, 2, 3, and 4 red marbles?

This histogram shows how many among 16 samples of 4 we'd expect (theoretically) to contain 0, 1, 2, 3, and 4 red marbles:

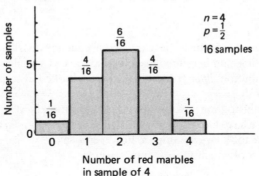

Number of red marbles in sample of 4

What if you took 32 samples of 4. How many of these would you expect (theoretically) to contain 0, 1, 2, 3, and 4 red marbles?

Given the probabilities:

0 Red	1 Red	2 Red	3 Red	4 Red
$\dfrac{1}{16}$	$\dfrac{4}{16}$	$\dfrac{6}{16}$	$\dfrac{4}{16}$	$\dfrac{1}{16}$

You can work out the relative frequency for any number of samples. With exactly 32 samples we'd expect the theoretical frequency distribution shown in the histogram overleaf:

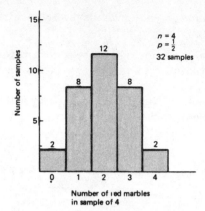

Now here is another binomial probability distribution you may remember from earlier on. It shows the probabilities of getting 0, 1, 2, 3, and 4 red marbles in a sample of 4 when $p(\text{red}) = \frac{1}{3}$.

0 Red	1 Red	2 Red	3 Red	4 Red
$\frac{16}{81}$	$\frac{32}{81}$	$\frac{24}{81}$	$\frac{8}{81}$	$\frac{8}{81}$

Draw a histogram to show the 'expected' frequency distribution of samples with each possible number of red marbles if exactly 81 samples are taken. How does the distribution *differ* from those we have looked at so far?

CHECK your histogram. (Notice that the distribution is NOT symmetrical— unlike those we have looked at so far. This is because p and q are *not equal* in this case. If we took larger samples, however—say $n = 20$—then the distribution would look much more symmetrical.)

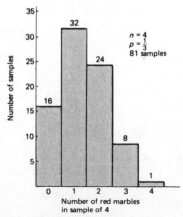

Remember, of course, that the expected frequency distribution is *theoretical*. It applies only 'on average' and 'in the long run'. Your chances of getting exactly what you expect from just one set of samples are extremely slim.

Do you remember the experiments in probability I suggested on pages 125–6. If you tried any of them, you should at this stage be able to compare your actual results with the theoretical results.

For instance, I suggested tossing 6 coins together 128 times and noting the number of heads in each set of six. *Work out* the theoretical frequency distribution for 128 samples of 6 and *draw* a histogram. Will the distribution be symmetrical?

The probabilities are:

No. of heads	0	1	2	3	4	5	6
Probability	$\dfrac{1}{64}$	$\dfrac{6}{64}$	$\dfrac{15}{64}$	$\dfrac{20}{64}$	$\dfrac{15}{64}$	$\dfrac{6}{64}$	$\dfrac{1}{64}$

So the expected frequency distribution over 128 samples is shown on the left below:

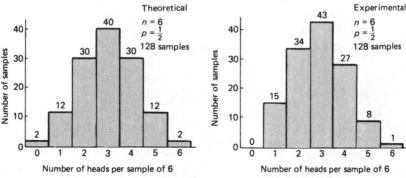

If you did the experiment, compare the theoretical distribution on the left with your actual results. If not, compare it with the histogram on the right, showing the results of an experiment of mine. Notice that my results follow the *symmetrical* pattern of the theoretical distribution (where $p = \dfrac{1}{2} = q$) and are fairly close to the expected results.

If I threw the 6 coins 1280 times instead of 128 only, would you think the expected results and actual results would be relatively further apart *or* would they be even closer together?

If I threw the 6 coins 1280 times instead of 128 times only, my theoretical and actual results would probably be even *closer together*. (The larger the number of samples, the better the chance for the expected distribution to develop.)

Now try this question: (Notice that it has *two* parts).

If three dice are thrown together a large number of times, the frequencies with which 0, 1, 2, 3 sixes are obtained may be taken as proportional to the terms of the expansion $(q + p)^3$ where $p = \dfrac{1}{6}$ and $q = \dfrac{5}{6}$. Find the expected frequency distribution if the dice are thrown 1080 times.

On how many occasions would you expect to obtain at least 2 sixes?

The probabilities with a sample of 3 are:

$$0 \text{ sixes} \ldots \frac{3!}{3!0!}q^3 = q^3 = \left(\frac{5}{6}\right)^3 = \frac{125}{216}$$

$$1 \text{ six} \ldots \frac{3!}{2!1!}q^2p = 3q^2p = 3\left(\frac{5}{6}\right)^2\left(\frac{1}{6}\right) = \frac{75}{216}$$

$$2 \text{ sixes} \ldots \frac{3!}{1!2!}qp^2 = 3qp^2 = 3\left(\frac{5}{6}\right)\left(\frac{1}{6}\right)^2 = \frac{15}{216}$$

$$3 \text{ sixes} \ldots \frac{3!}{0!3!}p^3 = p^3 = \left(\frac{1}{6}\right)^3 = \frac{1}{216}$$

So the expected frequency distribution of 0, 1, 2, 3 sixes in 216 samples would be 125, 75, 15 and 1. Since we are asked to consider five times as many samples (1080 = 216 × 5), we would expect each frequency to be five times as large— *625, 375, 75, and 5*.

In 75 of these 1080 samples we'll have thrown 2 sixes and in 5 of them we'll have thrown 3 sixes. So we'll have obtained 'at least 2' sixes (i.e. 2 or 3) on 75 + 5 = *80 occasions*.

Were you right on both parts of the question? (Check through anything you are not clear about.)

Samples and Populations

The binomial distribution is used a great deal in scientific work and in industry, where manufactured goods have to be inspected by *sampling*. Why are manufactured goods inspected? To reject faulty articles and make sure that as few duds as possible are passed on to the customer.

Why then, are the articles merely sampled? Why is each article not inspected individually for faults? Three reasons.

1   To inspect every item might add so much to the cost of manufacture that it could no longer sell at a price customers could afford to pay.
2   Even if each item were inspected individually, human oversight would ensure that some duds still slipped through.
3   Some inspections involve tests that destroy the article (e.g. electric light bulbs are tested by connecting them to a supply of electricity and stepping up the current until the element gives out); 100% testing here would leave the manufacturer with nothing to sell.

So sampling inspection is concerned with the proportion of faulty articles being handed on to the customer. How does sampling aim to affect this proportion of faulty goods?

(a) By keeping it to an acceptable minimum?

or (b) By eliminating it altogether?

Sampling inspection aims to *keep to an acceptable minimum (a)* the proportion of faulty goods being passed on to the customer.

With mass-produced articles it would be foolish to expect 100% perfection. (And 99.9% perfection in an automobile with 20 000 components means 20 defective components.) So the purpose of sampling inspection is to give warning if the proportion of defective goods being produced rises above a permissible level—let's say $2^1/_2\%$. Samples of 10, 20, or more articles are checked as they come off the production line, and the proportion of duds in the samples is taken as an indication of the proportion in the production-run from which they were drawn.

But this is not necessarily as easy as it sounds. Let's say we are manufacturing ball-point pens in huge quantities. We take a sample of 4 from a production run, and the sample contains one defective pen.

What proportion of defective pens would you expect to find if you examined all the production run?

(a) 25%?

or (b) Less than 25%?

or (c) More than 25%?

or (d) We really can't tell?

We really can't tell (d). This is the wisest answer. From such a small sample it would be dangerous to jump to any conclusions about the population as a whole. The production run might show the same 1 in 4 ratio as the sample we took from it, but this is very far from certain.

In actual fact, the defective proportion in that production run was 10%. Then how come you get a sample with 25% duds? Easy. Consider the binomial probability distribution with $n = 4$ and $p = \dfrac{1}{10}$. The probabilities of drawing 0, 1, 2, 3, 4 duds should be equal to the expansion of $\left(\dfrac{9}{10} + \dfrac{1}{10}\right)^4$.

You should have no trouble working this out, so:
complete the table

No. of duds in $n = 4$	0	1	2	3	4
Probability					

CHECK:

No. of duds in $n = 4$	0	1	2	3	4
Probability	$\dfrac{6561}{10\,000}$	$\dfrac{2916}{10\,000}$	$\dfrac{486}{10\,000}$	$\dfrac{36}{10\,000}$	$\dfrac{1}{10\,000}$

And as you can work out from these figures:

65.6% of samples taken would suggest that there were NO duds at all in the production run as a whole.

29.2% of samples would suggest the production run was 25% defective.

4.9% of samples would suggest 50% of the run was defective; and about $\frac{1}{3}$% would suggest a defective rate of 75% or 100%.

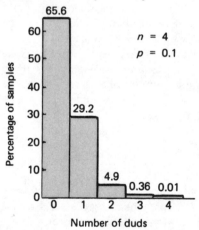

Number of duds

Thus our sample of 4 with one defective was nothing unusual—we could expect about 29 such samples in every hundred we took.

This example shows up the danger of relying on a single small sample. After all, two thirds of these samples suggest that the production run contains (*more/fewer?*) defectives than it actually does.

... contains *fewer* defectives than it actually does.

One way of overcoming this, and getting a more reliable estimate of the number of duds in the production run, is to take *larger* samples. Here, for instance, is the expected frequency distribution for $n = 20$ (and $p = \frac{1}{10}$) based on the expansion of $\left(\dfrac{9}{10} + \dfrac{1}{10}\right)^{20}$

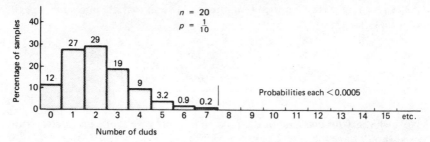

Number of duds

Thus, if we are taking samples of 20, only about (*how many?*) samples in every hundred will suggest that the production run contains less than 10% defectives.

... only about *39* (i.e. 12 + 27) samples in every hundred will suggest that the production run contains less than 10% defectives. (Compared with more than 65 in every hundred with samples of 4.) So if you have to rely on a single sample, the bigger the better. But even with a sample of $n = 20$, as in that last example, you'd have only a 29% chance of choosing a sample with *exactly* the same proportions as the whole run.

In fact, if the samples have to be small ones you can only make up for it by taking lots of them. For example, we have taken a number of samples of 4 from our production run of ball-point pens, and this is how the samples are distributed:

No. of duds in sample	0	1	2	3	4
No. of samples of $n = 4$	16	7	2	0	0

Total: 25 samples.

(i)   How many pens were tested altogether?
(ii) How many defective pens were found?
(iii) What proportion of the pens tested were defective?
(iv) What proportion of defectives would this suggest in the production run as a whole?
(v)   How could we improve on this estimate?

(i)   Altogether $25 \times 4 = 100$ pens were tested.
(ii) Of these, $(0 \times 16) + (1 \times 7) + (2 \times 2) = 11$ were duds.
(iii) So $\frac{11}{100}$ of the pens tested were defective.
(iv) This suggests that 11% of the production run was defective.
(v)   The only way we can get a better estimate is by taking more samples—that is, increasing the number of pens tested.

So let's say that we test another 75 samples of 4 each, so that our results overall now look like this:

No. of duds in sample	0	1	2	3	4
No. of samples of $n = 4$	64	32	3	1	0

Total: 100 samples.

Now what is your estimate of the proportion defective in the production run as a whole?

Your estimate should be reasoned as follows:

(i)   100 samples of 4 were tested—400 in all.

(ii) Of these $(0 \times 64) + (1 \times 32) + (2 \times 3) + (3 \times 1) + (4 \times 0) = 41$ were duds.

(iii) The samples contain a defective proportion of $\dfrac{41}{400}$.

(iv) So our best estimate of the proportion defective in the whole production run is $\dfrac{41}{400} \times 100 = 10.25\%$.

If we went on and doubled the size of the overall sample we might get closer still to the true proportion.

Try another example:

A manufacturer packs table-tennis balls in boxes of one dozen. On testing 50 boxes about to leave the factory, he gets these results:

No. of defectives in box	0	1	2	3	4	5 or more
No. of boxes	18	19	9	3	1	0

What is the probability that a ball chosen at random from the production run as a whole will be defective?

(i)   The number of balls tested is $50 \times 12 = 600$.

(ii) Of these $(1 \times 19) + (2 \times 9) + (3 \times 3) + (4 \times 1) = 50$ are defective.

(iii) That is, $\dfrac{50}{600} = \dfrac{1}{12}$ of the total sample are defective.

(iv) So the probability of choosing a defective ball from the production run as a whole is $\dfrac{1}{12}$ or $8\dfrac{1}{3}\%$.

Review Questions

1   The probability that a man aged 30 years will die before the age of 60 is $\dfrac{1}{4}$.

What is the probability that, of a group of eight men aged 30, exactly four will be alive to attend a reunion in 30 years time?

2 I am going to Seattle for a week and am told it rains three days out of four there. What is the probability it will rain on no more than two days out of the seven I shall spend in Seattle? (Assume the days are independent of one another.)

3 The probability that a rocket will fire successfully is 0.7. If a test series involves firing twelve of them, what is the probability that at least nine will fire successfully? (Leave your answer in nCr form.)

4 If five pennies are tossed simultaneously, what will be the ratio of the frequencies with which 0, 1, 2, 3, 4, or 5 heads appear (theoretically) in a large number of samples?

Answers to Review Questions

1 $(8\text{C}4)q^4p^4 = 70\left(\dfrac{3}{4}\right)^4\left(\dfrac{1}{4}\right)^4 = \dfrac{5670}{65\,536} = 0.0865$

2 $(7\text{C}2)q^5p^2 + (7\text{C}1)q^6p + 7\text{C}1\ q^7 =$
$$15\left(\frac{1}{4}\right)^5\left(\frac{3}{4}\right)^2 + 7\left(\frac{1}{4}\right)^6\left(\frac{3}{4}\right) + 1\left(\frac{1}{4}\right)^7 = \frac{157}{16\,384} = 0.0096$$

3 $12\text{C}9(0.3)^3(0.7)^9 + 12\text{C}10(0.3)^2(0.7)^{10} +$
$$12\text{C}11(0.3)(0.7)^{11} + 12\text{C}12(0.7)^{12}$$

4 $1 : 5 : 10 : 10 : 5 : 1$

Review Questions (Continued)

5 Families with 6 children may have 0, 1, 2, 3, 4, 5, or 6 girls. Write down an expected frequency distribution if 320 such families are chosen at random.

6 Earlier, I suggested you draw at least 128 samples of 4 from a bag of marbles, 25% of which were of a special colour, say red. Draw up an expected frequency distribution for 128 such samples. What proportion of samples would you expect to contain exactly 25% red marbles?

7 I have a box containing blue and yellow marbles. I take 162 samples of 4 marbles, getting the following distribution:

No. of blues	0	1	2	3	4
No. of samples	30	65	52	14	1

If the bag contains exactly 480 marbles, about how many would you say are blue? Calculate the theoretical binomial distribution for 162 samples of 4 based on your estimate of the probability of choosing a blue marble from the bag.

Answers (Continued)

5 No. of girls	0	1	2	3	4	5	6
No. of families	5	30	75	100	75	30	5

6 No. of reds   0   1   2   3   4
    No. of samples 41 54 27   6   0

(That is, according to the ratio $81 : 108 : 54 : 12 : 1$. And about $\dfrac{12}{256}$ of all samples should contain exactly 3 red marbles.)

7 Out of $162 \times 4 = 648$ marbles, 215 are blue; so the proportion of blue marbles in the samples is about $\dfrac{1}{3}$. This would suggest that about 160 of the 480 marbles in the box are blue.

Assuming the probability of choosing a blue marble really is $\dfrac{1}{3}$, the theoretical frequency distribution for 162 samples of 4 is:

	No. of blues	0	1	2	3	4
	No. of samples	32	64	48	16	2

Review

The subject of probability is far from exhausted. (If it were relevant, and if you were already familiar with the statistical notions of mean, standard deviation and normal distribution, we could easily go on to discover that, with large samples, the shape of a binomial distribution approaches that of the normal curve. We could then calculate the mean and standard deviation of a binomial distribution and use them, together with the constant proportions of the normal distribution, to estimate probabilities that would involve you in endless arithmetic were you to try calculating them by the methods you've used so far, e.g. like the probability of getting exactly 30 heads among 100 tossed coins. If you want to follow this up on your own, look at *Principles of Statistical Techniques* (Chapter 9) by P. G. Moore (published by Cambridge University Press 1976).

Meanwhile, ... we have covered a great deal of ground in Chapters 3 and 4. To remind you of the main ideas, read once more through the following review panels from earlier sections. You should *understand* them all and be able to *use* the ideas in them. Make sure you have a copy of each one.

If there are a ways of doing one thing, b ways of doing a second thing, c ways of doing a third (and so on), then the number of ways of doing all these things is $a \times b \times c$ (and so on).

The number of PERMUTATIONS of n different items taken r at a time (without repetitions) is

$$n\mathbf{P}r = \frac{n\mathbf{P}n}{(n - r)!} = \frac{n!}{(n - r)!}$$

If you have n things, containing a alike of one kind, b alike of a second kind, c alike of a third kind (and so on ...); then the number of PERMUTATIONS of the n things taken all together is:

$$\frac{n!}{a!b!c! \ldots}$$

The number of COMBINATIONS from a set of n different items, taken r at a time, is

$$n\mathbf{C}r = \frac{n\mathbf{P}r}{r!} = \frac{n!}{(n - r)!r!}$$

Problem

What is the probability of selecting a sample containing so many items of type A and so many of type B (e.g. men and women, Russians and Americans, Corvettes and others, etc.) from a population containing a known number of each type?

Combinations Method

① In how many ways can we select a sample of the required total *size* from the population?

② In how many ways can we select the required number of type A items?

③ In how many ways can we select the required number of type B items? (And we could continue to types C, D, E, etc.).

④ Multiply ② and ③ to give total numbers of ways we can select a sample of required proportions

⑤ Divide by ① to find the probability of getting that sample.

$$\frac{② \times ③}{①} = p$$

If an item (e.g. a baby) can take either of two forms (e.g. boy or girl) whose probabilities are p and q, then the probabilities that a sample of n such items will contain 0, 1, 2, 3, ... $n - 3$, $n - 2$, $n - 1$, n boys are given by the successive terms of $(q + p)^n$.

If you want to calculate the probability of getting r of one kind of thing in a sample of n things, the coefficient for the probability term $q^{n-r}p^r$ is $n\mathbf{C}r$.
That is, Probability $= (n\mathbf{C}r)q^{n-r}p^r$

$$= \frac{n!}{(n-r)!\,r!}q^{n-r}p^r$$

Practice

Of course, it is not enough merely to understand and be able to use the techniques we've discussed *now*. To keep your skill in problem-solving and become really confident, you must *practise* the techniques on as wide a variety of problems as you can find.

Here now, to complete the book, is a final set of probability problems. They are a mixed bunch. Some are much more difficult than others and, between them, they should enable you to practice what you have learned from all chapters of this book. The answers follow immediately afterwards.

Examination Questions

1 The probability that a student attempts one particular question in an examination is $\frac{9}{10}$ and, having done so, the probability of his or her success is $\frac{2}{3}$. What is the probability that the examiner will find at least one correct solution in the first three exams which he marks.

2 Calculate the probability of three or more successes in four independent binomial trials in each of which the probability of success is $\frac{1}{4}$.

3 The players trying out for the hockey team consist of 11 forwards, 4 goalies and 8 defensemen. In how many different ways can a team consisting of 8 forwards, 2 goalies and 5 defensivemen be selected?

4 If, after the team in Question 3 has been selected. 1 forward, 1 goalie and 2 defensemen are found to be unable to play, what is the probability that the team will not have to be changed?

5 It is estimated that one in ten people who holiday go to the beach. A pupil interviewed a random sample of 3 people who intended to go on a holiday. Calculate the probabilities that:
(i) all three people interviewed intended to go to the beach;
(ii) of the three people interviewed, two or more intended to go to the beach.
If 100 such random samples were investigated, in how many of these would the pupil expect to find *fewer* than two people who intended going to the beach on holiday?

6 A die is thrown ten times. Which of these two results has the higher probability: just one six, or just two sixes?

7 At a village soccer match the probability that each potential player turns out to play is 0.9 independently of the other potential players. The home side has a goal keeper and 11 other players available. Find the probability that the home side will have at least a complete team (which consists of a goal keeper and 10 other players) present on the day of the match.

The away side has 2 goal keepers, who can also play outside the goal, and 11 other players. Find the probability that the away team will be at least complete on the day of the match.

If one team is complete and the other is not, then the complete team wins by default. Find the probability that (a) both teams are complete, (b) the away team wins by default. $[(0.9)^{11} = 0.3138]$

8 At a pottery, when running normally, 20% of the teacups made are defective. Find the probability that a sample of five will contain:
(i) no defective;
(ii) exactly one defective;
(iii) at least two defectives.

9   (a) Two unbiased, six-sided dice are thrown. What is the probability that
(i)   the total score is 12,
(ii) the total score is 4,
(iii) both dice show the same score?
(b) Random numbers are generated in such a way that, in each position, each of the ten digits from 0 to 9 has an equal chance of appearing. If random numbers are grouped into blocks of four digits, what is the probability that in a particular block
(i)   all four digits are the same,
(ii) all four digits are different,
(iii) three digits are the same and one is different,
(iv) two or more digits are the same?

10   (a) The outcome of an experiment may be any one of four mutually exclusive events. The respective probabilities of three of the events are 0.1, 0.3 and 0.2. Calculate the probability that the experiment results in the fourth event.
(b) An unbiased die, marked with the numbers 1, 3, 3, 3, 4, 4, is thrown twice. What is the probability that both throws result in a 3?
(c) In a typhoid epidemic the following data was collected:

	Infected	Not infected
Inoculated	500	10 000
Not inoculated	250	1 000

A society of 'anti-inoculators' used these figures to argue that since twice as many inoculated people were infected, inoculation was more dangerous. Explain briefly the fallacy in this argument.

11   (a) In backgammon use is sometimes made of a six-sided die numbered 2, 4, 8, 16, 32, 64. If this die and a conventional die (numbered 1, 2, 3, 4, 5, 6) are thrown together, what is the probability that
(i)   the total score will be 3,
(ii)   the total score will be 6,
(iii) the total score will be 16,
(iv)   the total score will be an even number,
(v)   the score on the backgammon die will be less than the score on the conventional die?
(b) A machine producing tablets has twelve heads, one of which is faulty. Thus, one twelfth of the tablets produced by the machine come from the faulty head. If a sample of three tablets is selected, what is the probability it will contain
(i)   no tablets from the faulty head,
(ii)   exactly one tablet from the faulty head,
(iii) at least one tablet from the faulty head?

12 A manufacturer of glass marbles produces equal numbers of red and blue marbles. These are thoroughly mixed together and then packed in packets of six marbles which are random samples from the mixture. Find the probability distribution of the number of red marbles in a packet purchased by a boy.

Two boys, Fred and Tom, each buy a packet of marbles. Fred prefers the red ones and Tom the blue ones, so they agree to exchange marbles as far as possible, in order that at least one of them will have six of the colour he prefers. Find the probabilities that, after exchange: (i) they will both have a set of six of the colour they prefer; (ii) Fred will have three or more blue ones.

13 (a) How many permutations of four cards can be taken from a standard pack of playing cards? Give the probability that, of four such cards selected randomly,

(i) exactly one card is an ace;

(ii) at least one card is an ace;

(iii) no two cards are of the same suit;

(iv) no two cards are of the same denomination.

(b) Two cards are dealt from a standard pack. What is the probability that the second card dealt is a diamond if the first one is

(i) also a diamond;

(ii) a spade?

Hence or otherwise find the probability that the second card dealt is a diamond.

14 Three girls, two of whom are sisters, and five boys, two of whom are brothers, meet to play tennis. They draw lots to determine how they should split up into two groups of four to play doubles.

(a) Calculate the probabilities that one of the two groups will consist of: (i) boys only; (ii) two boys and two girls; (iii) the two brothers and the two sisters.

(b) If the lottery is organised so as to ensure that one of the two groups consists of two boys and two girls, calculate the probability that the two brothers and the two sisters will be in the same group. Given that the two brothers are in the same group, calculate the probability that the two sisters are also in that group.

15 The probability of damaging any corner of a square tile during manufacture is 0.1, independently of the other corners. Find

(a) the proportion of tiles that are completely undamaged,

(b) the proportion with one corner damaged,

(c) the proportion with a pair of adjacent corners damaged.

Undamaged tiles are worth 10¢ each, while those with one corner damaged are worth 2¢ each and those with two adjacent corners damaged are worth 1¢ each. The others are worthless. Find the expected value of 1000 tiles.

16 In the game of backgammon you are not able to move until you have thrown a six on the die. Calculate the probability that you will have to throw the die more than three times before this occurs.

17   In a selected group of 100 people, 60 have brown hair and 45 have blue eyes. Find the chance that a particular member of the group has:
    (a) brown hair and blue eyes,
    (b) brown hair and eyes which are not blue,
    (c) neither brown hair nor blue eyes.

18   Three normal dice are thrown. Find the probability that: (i) the sum of the scores is 18; (ii) the sum of the scores is 5; (iii) none of the three dice shows a 6; (iv) the product of the scores is 90.

19   (a) In a lottery there are 900 tickets numbered consecutively 100, 101, ... , 998, 999. If a ticket is drawn at random, what is the probability that
    (i)   all three digits are the same,
    (ii) all three digits are different,
    (iii) exactly two of the digits are the same,
    (iv) the three digits are neither in ascending nor descending consecutive order (e.g. not of the form 345 or 765),
    (v) the three digits read the same backwards and forwards?
    (b) In a car race the probability that a certain driver will win is 0.6 if the weather is fine but only 0.3 if it is wet. If the probability that the race takes place on a wet day is 0.2, what is the driver's probability of winning?

(You may remember tackling parts (a) and (b) of this next and final question back in Chapter 3; but don't let that stop you trying it again!)

20   In how many ways can 5 unlike objects be arranged in order?

A party of 3 ladies and 2 gentlemen goes to a theatre and sits in a row of 5 seats in random order. Find the probability that
    (a) the two gentlemen sit together,
    (b) the ladies and gentlemen occupy alternate seats.

During the interval, three of the party get up and when they return each chooses one of the empty seats at random. Find the probability that (c) all three, (d) two, (e) one, (f) none of them, sit in the same seats as before.

Each of the five buys a ticket in a raffle for two prizes, in which 100 tickets are sold in all. These are put in a hat and the owners of the first two tickets drawn out receive the prizes. Find the probability that exactly one of the five wins a prize. (Answers may be given as fractions in their lowest terms.)

Answers to Examination Questions

1 $\dfrac{117}{125} = 0.936$

2 $\dfrac{13}{256} = 0.05$

3 55 440

4 0.98

5 (i) $\dfrac{1}{1000}$ (ii) $\dfrac{28}{1000}$; 97   **6** one 6

7 0.5963 0.6947 0.4142 0.1672

8 (i) 0.328. (ii) 0.41 (iii) 0.262

9 (a) (i) $\dfrac{1}{36}$ (ii) $\dfrac{1}{12}$ (iii) $\dfrac{1}{6}$

  (b) (i) $\dfrac{1}{1000}$ (ii) $\dfrac{504}{1000}$ (iii) $\dfrac{36}{1000}$ (iv) $\dfrac{468}{1000}$

10 (a) 0.4 (b) $\dfrac{1}{4}$; eight times as many people were inoculated, but only twice as many became infected. Their chance of becoming infected was only $\dfrac{1}{20}$, compared with $\dfrac{1}{5}$ if not inoculated.

11 (a) (i) $\dfrac{1}{36}$ (ii) $\dfrac{1}{18}$ (iii) 0 (iv) $\dfrac{1}{2}$ (v) $\dfrac{1}{6}$

  (b) (i) 0.77 (ii) 0.21 (iii) 0.23

12 Reds:

0	1	2	3	4	5	6
$\dfrac{1}{64}$	$\dfrac{6}{64}$	$\dfrac{15}{64}$	$\dfrac{20}{64}$	$\dfrac{15}{64}$	$\dfrac{6}{64}$	$\dfrac{1}{64}$

(i) 0.226 (ii) 0.073

13 (a) 6 497 400 (i) 0.064 (ii) 0.281 (iii) 0.105 (iv) 0.642

  (b) (i) $\dfrac{12}{51}$ (ii) $\dfrac{13}{51}$; $\dfrac{13 + 13 + 13 + 12}{51 + 51 + 51 + 51} = 0.25$

14 (a) (i) $\dfrac{1}{7}$ (ii) $\dfrac{6}{7}$ (iii) $\dfrac{1}{35}$ (b) $\dfrac{1}{30}, \dfrac{1}{12}$

15 (a) 0.6561 (b) 0.2916 (c) 0.0324; £71.76

16 $\dfrac{125}{216} = 0.579$

17 (a) 0.27 (b) 0.33 (c) 0.22

18 (i) $\dfrac{1}{216}$ (ii) $\dfrac{1}{36}$ (iii) $\dfrac{125}{216}$ (iv) $\dfrac{1}{36}$

19 (a) (i) $\dfrac{1}{100}$ (ii) $\dfrac{72}{100}$ (iii) $\dfrac{9}{100}$ (iv) $\dfrac{875}{900}$ (v) $\dfrac{1}{9}$

  (b) $(0.2 \times 0.3) + (0.8 \times 0.6) = 0.54$

20 120 (a) $\dfrac{2}{5}$ (b) $\dfrac{1}{10}$ (c) $\dfrac{1}{6}$ (d) $\dfrac{1}{6}$ (e) $\dfrac{2}{3}$ (f) $\dfrac{1}{3}; \dfrac{17}{198}$